Samuel S. Kingdon

Kingdon's Dictionary of the White Mountains

And Other New-England Summer Resorts

Samuel S. Kingdon

Kingdon's Dictionary of the White Mountains
And Other New-England Summer Resorts

ISBN/EAN: 9783337134167

Printed in Europe, USA, Canada, Australia, Japan

Cover: Foto ©Andreas Hilbeck / pixelio.de

More available books at **www.hansebooks.com**

KINGDON'S DICTIONARY

OF THE

WHITE MOUNTAINS

AND OTHER

NEW-ENGLAND SUMMER RESORTS.

BOSTON:
PUBLISHED BY THE AUTHOR.
1894.

Presswork by
JOHN WILSON AND SON, CAMBRIDGE, U.S.A.

INTRODUCTION.

The aim in the preparation of this book has been to present the information it contains in a convenient and concise form for ready reference. and in such a manner as to secure the greatest possible facility for consultation. All superfluous and unimportant matter has been studiously excluded, thus securing economy of space and price without omitting material facts.

Necessarily many interesting and attractive resorts have been omitted, for to include all the places in New England where a summer vacation may be pleasantly and profitably passed would be to mention nearly every town and hamlet within the territory included in this book. It is believed that no place of popular resort has been overlooked, and that all their principal attractions have been referred to. No mention of hotels has been made, but the tourist will find excellent accommodations at all the principal places. No statement, endorsement, or recommendation in this book has been influenced by any one, and the advertising matter has been confined to its proper place at the end of the book.

While much care has been taken to make this Dictionary accurate and trustworthy, it is probable that some errors may have crept in, and the Editor will be grateful for any corrections and suggestions he may receive.   S. S. KINGDON.

SUMMER RESORTS.

A

AGASSIZ BASIN. See North Woodstock.

ALBANY BASINS. See Bethel.

ALTAR, THE. A strange-looking stone of large size and unusual form on Mt. Lafayette. [See Mt. Lafayette]. The name it bears is derived from a fancied resemblance to the old Runic remains of a similar character.

ALTON BAY, N. H., a village on the Northern Division of the Boston & Maine Railroad, at the most southern point of Lake Winnipesaukee, is situated at the head of a narrow estuary, which appears more like a river than a lake. The Adventists hold campmeetings there. From Sheep Mountain, two miles north, there is a fine view of the lake; also from Prospect Hill and Mt. Major. Lougee Pond, about seven miles distant, is noted for its tame fish. The most interesting excursion is to the summit of Mt. Belknap, 10 miles distant. The view from the summit, 2062 feet above the sea, is very fine.

AMHERST, N. H. A pleasant rural town on the Keene branch of the Southern Division of the Boston & Maine Railroad, 48 miles from Boston. Milford Springs, one and a half miles from the station, have a reputation for their curative properties.

AMMONOOSUC FALLS. See Fabyans.

ANDOVER, N. H. A village on the Connecticut River Division of the Boston & Maine Railroad, 104 miles from Boston. In the vicinity are Eagle Pond, four miles long, and Ragged Mountain, overlooking it and the course of Blackwater River.

APPALACHIAN CASCADE. See Jackson.

APPLEDORE ISLAND. See Isle of Shoals.

ARTISTS' FALLS. See North Conway.

ASQUAM LAKES, popularly known as the Great and Little Squam, but restored to their ancient name of Asquam, with their lesser sister, Minnisquam, the largest of which is six miles long and three miles across at its broadest part, lie among the southern foot-hills of the White Mountains, half a dozen miles to the northeast of Lake Winnipesaukee, and about four miles east of Ashland, a pretty town on the Concord & Montreal Railroad. They are beautiful bodies of water, dotted with numerous small islands, and having shores irregular in form and picturesque in appearance.

B

BABY TWINS. See Twin Mountain Station.

BALD MOUNTAIN. One of the Franconia Range, 2310 feet high, from the summit of which a fine view can be had. The top is reached by a comparatively easy walk over a disused carriage road of a mile and a half from the Franconia Notch. [See Franconia Notch.] For another mountain of this name, see Lenox.

BAR HARBOR, on the eastern shore of Mt. Desert [See Mt. Desert], and just opposite the Porcupine Islands, derives its name from a sandy bar, visible only at low tide, which connects Mt. Desert with the largest and northermost of the Porcupine group. It is a popular and fashionable resort, on account of the fine scenery, the boating, and the fishing. Excursions can be made in carriages or on foot to the summit of Green Mountain, 1702 feet high, from which can be obtained a view embracing the whole of the island, Frenchman's Bay, with its many islands, and the ocean on the one hand, and a vast stretch of the Maine coast on the other. It is said that Katahdin, 100 miles distant, and Mt. Washington, 140 miles away, can be seen from this point. Eagle Lake is visible at intervals during the ascent, and half way up a short detour will bring the tourist to it. Mt. Newport is ascended from the Schooner Head road, and Kebo, the summit of which may be reached in half an hour, affords a fine prospect. A pleasant drive of seven miles through the woods will bring one to the Ovens, a series of cavities worn in the cliffs by the action of the tide, some of which are large enough to contain 30 or 40 people. They can only be visited at low tide. The Via Mala is a curious archway in one of the projecting cliffs. Schooner Head, so named from the resemblance that a mass of white rock on its sea face bears to a small schooner, is on the seaward side of the island, four miles south of Bar Harbor. The Spouting Horn is a wide chasm in the cliff, which extends down to the water, and opens to the sea through a small archway below high-water mark. At high tide, and especially in stormy weather, the waves rush through this archway, and send a spout of water far above the summit of the cliff. Great Head, two miles south of Schooner Head, is the highest headland between Cape Cod and New Brunswick. It is a bold projecting mass, the base of which has been deeply gashed by the waves. The best view of its front is obtained by descending to the foot of the cliff. Farther south are the Otter Creek Cliffs, situated near Otter Creek, a small stream. The most interesting feature of these cliffs is Thunder Cave, reached from the road by an excellent path through the forest. This is a long, low gallery, into which the waves rush with great force, producing a sound closely resembling thunder. Near it is the Obelisk, a tall, pointed column, with an apparently artificial base of steps, bearing a close resemblance to a monument of stone. On the cliffs to the westward is Castle Head, the wall of which looks like the ruins of a castle. About nine miles southwest of Bar Harbor is Jordan's Pond, a beautiful lake two miles long and half a mile wide, surrounded by picturesque mountain scenery, and abounding in fish. Cromwell's Cave, the Pulpit, the Indian's Foot, and the Assyrian, a rock figure in one of the cliff sides, are in this vicinity.

BARNES' FALLS. See Wilton.

BARTLETT, N. H., noted for its magnificent surroundings, is on the main road leading to the Crawford Notch [see Crawford Notch], and forms an important station of the Maine Central Railroad. It is situated on the Saco River, and is hemmed in by grand mountains.

BASIN, THE. A beautiful freak of nature near the roadside in the Franconia Notch [see Franconia Notch], about a mile from the southerly entrance. Here the waters of the Pemigewasset fall over a rocky ledge, a few feet in height, into a deep hollow in the solid granite, formed by the continual action of the water and mingled stones and bowlders from above. The diameter of the basin is about 30 feet in its shortest width, and 40 feet in its longest. Its circumference is about 60 feet, and its depth ordinarily about 15 feet. It is filled at all seasons with cold, pure, and pellucid water, through which the bottom of the basin can be distinctly seen. The waters, in escaping over the side, form a series of beautiful waterfalls.

BASS ROCK, 30 miles from Boston, on the Eastern Division of the Boston & Maine Railroad, is situated on the high rocky shore of Gloucester, Mass., between Eastern Point and Rockport. It commands an extensive view of land and sea. Good Harbor Beach, three quarters of a mile in length, is the finest on the North Shore for surf bathing, and is safe at any time of tide. There is also a shallow inlet, with clean sand bottom, which is a favorite place for those who prefer still-water bathing. There are many fine drives, and sailing and deep-sea fishing are favorite pastimes.

BAY OF NAPLES. See Sebago Lake.

BEACH BLUFFS. See Swampscott.

BEACH MOUNTAIN. See Southwest Harbor.

BEAMIS POND. A small sheet of water in Crawford Notch [see Crawford Notch], the source of Sawyer's River. It was formerly a favorite resort for trout fishers.

BEAR RIVER NOTCH. See Grafton Notch.

BERLIN FALLS, six miles from Gorham [see Gorham], are reached by a carriage road along the course of the Androscoggin River. At this point for a course of a mile the river descends nearly 200 feet, swift, rapid, and broken here and there by a direct and powerful fall, and by some it is regarded as one of the most interesting falls in the country.

BETHEL, ME., a pretty town on the Grand Trunk Railway, 21 miles from Gorham, has many places of interest. On one side, 12 miles distant, are the Albany Basins, worn out of the solid granite, and, on the other, 18 miles distant, are the Rumford Falls, where the Androscoggin River makes a descent of 160 feet in three pitches, and within the space of a quarter of a mile. There is one sheer descent of 70 feet. Fifteen miles from Bethel are Screw Auger Falls.

BETHLEHEM, N. H. A beautiful village, 201 miles from Boston, famous for its view of the whole range of the White Mountains. It was first settled in 1790, under the name of Lord's Hill. In 1803 the town consisted only of a few log huts, and to-day it is the site of some of the largest and most famous hotels in the world. Its elevation, 1489 feet, is claimed to be greater than that of any town in New England, and no town in the mountain region has made such marked and rapid growth. With the Franconia Range on the one hand and the Presidential Range upon the other, the chief mountain outlooks are grand in the extreme, while every principal point is easily accessible. Besides the wonderful view of the White Mountains to be had from any part of the village, in the northwest are to be seen some noble white marble mountains in Vermont which have the appearance of being covered with snow. From Mt. Agassiz, which is a little southeast of the village, and 2042 feet high, and also from many lesser elevations, wide-sweeping views may be had. The exemption of the town from hay fever has made it a famous resort for invalids. The drives from here to Franconia, Littleton, and other points of interest, are through the most grand and diversified scenery in the mountains, and add greatly to the attractions of the village as a summer home. A spur of the Profile & Franconia Notch Railroad ascends the hill, and traverses the outskirts of the village from one end to the other, connecting at Bethlehem Junction with the White Mountain branch of the Concord & Montreal Railroad.

BEVERLY FARMS, a favorite sea-side home of citizens of Boston, who own beautiful residence there, is on the Eastern Division of the Boston & Maine Railroad, 20 miles from Boston. The beaches are fine, and from there to Manchester-by-the-Sea are to be seen some of the best examples of landscape gardening in the country.

BOAR'S HEAD. See Hampton Beach.

BOSTON & MAINE RAILROAD. The direct connect-

ing link between Boston and the PresidentialRange
of the White Mountains. For 110 miles, or a little
more than four hours' ride, the journey is made
through some of the most lovely and picturesque
landscapes in Massachusetts and New Hampshire.
From Wolfeboro Centre, the first mountain station
on the road, a branch road, about 12 miles long,
runs to Wolfeboro, where close connection is made
with the line boats on Lake Winnipesaukee. One
desiring to ride the entire length of the lake should
take the branch road at Rochester, N. H., for Alton
Bay. [See Wolfeboro and Alton Bay]. In this
way the whole western side of the mountains, including the Franconia Range, can be reached, and
the tour be extended over an almost unlimited and
inexhaustible territory. An hour's ride on the
main road from Wolfeboro Junction, with the
mountains continually in sight, and Conway, the
southern entrance to the great Crawford Notch, is
reached. Five miles beyond is North Conway,
where the cars pass on to the tracks of the Maine
Central Railroad, and at Bartlett, 10 miles beyond,
passengers are transferred to observation cars,
with no additional expense, from which they can
view uninterruptedly the grandeur and wonders
of the Notch. [See Crawford Notch]. The Eastern Division of the road is the direct route by rail
to the resorts along the North Shore, including the
Isles of Shoals, Wells Beach, Old Orchard, and Bar
Harbor.

BRENTON'S COVE. See Newport, R. I.

BRENTON'S POINT. See Newport, R. I.

BRIDAL VEIL FALLS. A beautiful cascade near
Franconia [see Franconia] of 75 feet descent, upon
Copper Mine Brook, which flows down the west-
ern slope of Mt. Kinsman.

BRIDGTON, ME., a thriving village lying about 40
miles northwest of Portland, is pleasantly situated
within a mile of the steamboat landing near the
head of Sebago Lake navigation. [See Sebago
Lake.] It was originally named Pondicherry, from
the number of ponds and cherry trees in the vicinity, and rechristened Bridgton after a person of
the name of Bridges. It is reached from Sebago,
a station on the Maine Central Railroad, by a delightful sail over the lake and the Songo River, or
by the Bridgton & Saco River Railroad from the
junction of the two roads at Bridgton Junction,

19miles beyond Sebago. The neighborhood possesses many pleasant walks and drives. The principal places of interest are Highland Lake, Dodge's
Hill, Sunset Rock, and Forest Avenue, near by,
and in the immediate vicinity, Mt. Pleasant [see
Mt. Pleasant], Long Lake, North Bridgton Acad-
emy, home and grave of "Artemas Ward," Summit Mineral Spring, and the boyhood home of
Nathaniel Hawthorne.

C

CALDRON CLIFF. See Nahant.
CAMEL'S HUMP, one of the principal peaks of the
Green Mountain Range [See Green Mountains], is
4188 feet high. It may be ascended without much
difficulty from two sides, but it is most conven-
iently visited from Ridley's Station, a small village
on the Vermont Central Railway, five miles below
Waterbury, Vt. [see Waterbury]. Carriages run
from the station to the summit of the mountain,
three miles, and the remainder of the ascent may
be made either on horseback or on foot. The
mountain is covered with jagged rocks, and the
imposing scene from the summit is in no way ob-
structed by trees or other obstacles. The view
closely resembles that from Mt. Mansfield [see Mt.
Mansfield], except that that noble peak now forms
one of the most striking features of the landscape.
The beautiful Bolton Falls are a little below Ridley's Station.
CAMPTON. A beautiful little village on the Pemigewasset Valley Railroad, seven miles from Plymouth, N. H., and 133 miles from Boston. It is
situated on the Mad River, two miles from its con-
fluence with the Pemigewasset, and is rapidly becoming a favorite resort with tourists, as it has
long been with sportsmen and artists It is said to
contain more points for fine prospects than any
place in the neighborhood. One of the prominent
objects seen while traversing this part of the route
is the shapely mass of Welch Mountain, which
rises to the height of 3500 feet, north of the valley.
Mt. Weetamo, 2546 feet high, is on the south side,
and the summits of both are visited on account of
their extended view. Up the valley Trypyramid
and Sandwich Dome are to be seen.
CAPE NEDDICK. See York Beach.

CARTER DOME. See Carter Notch.

CARTER MOUNTAIN. See Carter Notch.

CARTER NOTCH. A wild and picturesque pass between Carter and Wild Cat mountains. It is about 10 miles from Jackson, N. H. [see Jackson], and is reached by following the carriage road up the Wild Cat River for six miles, and a foot path through the woods on the right for four miles. The path is very well defined, but it is tortuous and rugged. It was built and is kept in repair by a hunter and guide named Jonathan Davis, who lives near the terminus of the carriage road, and exacts a fee of 25 cents for its use. The Notch is remarkable as a scene of grandeur and desolation. Massive rocks 20 feet high are piled promiscuously in the valley formed by the two mountains, and they appear to have been thrown there by some great convulsion of nature. In the midst of the Notch is a small pond which serves but to add to the loneliness of the scene. Wild Cat Mountain rises abruptly from the valley 4350 feet, and Carter Mountain and Dome on the opposite side, 4830 and 4702 feet. From the top of one of the rocks, called The Pulpit, reached by a rude stairway, a fine view of the entire Notch can be had. There is a rude camp with a bed, stove, and cooking utensils, where parties are in the habit of spending the night; the round trip, unless carriages or horses are used the first six miles, being a difficult feat to accomplish in one day. There is a foot path from The Glen [see The Glen] to the Notch which is about two miles in length.

CASCADE, THE. A continuous fall of water from Flume Brook [see Flume Brook,] of more than 600 feet, with a gradual descent, but occasionally quite abrupt. It is near the southerly end of the Franconia Notch [see Franconia Notch], about three-quarters of a mile from the main road, in a south-easterly direction. A good carriage road leads to the lower part of the Cascade, and a footpath leads up the course of the stream.

CATSUPTIC LAKE. See Rangeley Lakes.

CENTRE HARBOR, a pretty town, ten miles distant from The Weirs by steamer, lies at the northern extremity or head of Lake Winnipesaukee, near 'Squam Lake [see Asquam Lake], and is 553 feet above the level of the sea. The location is a beautiful one and has always been deservedly popular.

The chief object of interest in the immediate vicinity is Red Hill. [See Red Hill.] Another place of almost equal interest is Shepard Hill, two miles distant on the carriage road between Centre Harbor and Ashland. It rises 800 feet in the centre of the land dividing the three Asquam lakes. The grandeur and beauty of the view from the summit is marvellous, embracing, as it does, not only the three lakes mentioned, but the most picturesque portion of Lake Winnipesaukee. It is the subject of Whittier's poem, "The Hill-Top." It was for years a favorite view of the Quaker poet, and he visited it every summer.

CHOCORUA ("The Old Bear"), one of the most notable of the lower peaks of the White Mountain Range. It is 3540 feet high, and without vegetation other than such as a few blueberries and cranberries create. It is noted for its steepness, the sharpness of its apex, and the extended view from its summit. It is, in fact, a granite mountain, with pinnacles and precipices, sharp, angular peaks, and unexpected descents; when viewed from certain positions, its top seems actually to overhang. It takes its name from an Indian chief who, tradition says, was shot on the summit by Cornelius Campbell, a settler, whose wife and children the Indian had murdered. The tradition says further, that the "Old Bear" with his dying breath, cursed the mountain, and a pestilence among the cattle, and other calamities, were formerly ascribed to its influence. The ascent, which is very difficult, can be made from Ossipee. [See Ossipee.]

CHOCORUA LAKE. A beautiful body of water at the foot of Chocorua Mountain.

CLAREMONT, N. H. A very pleasantly situated town on the Connecticut River, 129 miles from Boston, on the Concord & Claremont Branch of the Concord & Montreal Railroad. The surrounding scenery is on a grand scale, Ascutney Mountain, across the river in Vermont, looming above in its grandeur and overlooking the town, while to the northeast Green and Bald mountains are only a part of the grander Croydon Mountain, a prominent feature of the landscape a little farther removed.

CLIFTON. See Swampscott.

COHASSET, a pleasant village on the rocky coast of

Massachusetts, is twenty-one and a half miles from Boston, via the Old Colony Division of the New York, New Haven and Hartford Railroad. The coast line here is extremely rugged and broken, but picturesque and romantic, and is lined with villas and hotels. Minot's Ledge Lighthouse is situated here. Two and a half miles distant, in the direction of Boston, is Nantasket, a popular seaside resort, with a beautiful beach four miles long, with steamboat connections with Boston.

COLD BROOK. See Randolph.

CONANICUT ISLAND. See Newport, R. I.

CONCORD, N. H., the capital of New Hampshire, is pleasantly situated on level and gradually rising land overlooking the wide intervales of the Merrimack River, 75 miles from Boston, on the Concord & Montreal Railroad. It was originally called Penacook, from an Indian tribe over which Passaconaway held sway, whose home was near by; afterwards called Rumford, then Bow, and, in 1765, Concord. Included in the city limits are East and West Concord. It is one of the most interesting inland cities of New England. Its typography presents a wonderful variety, consisting of hills and wide-spreading intervales, winding rivers, swift rapids, and calmly nestling lakes. Through the centre of the city north flows the tortuous Merrimack, and the Contoocook enters from the northwest corner. The hills of Concord would in many states be designated as mountains, for one — the Rattlesnake — rises from near the centre of the city to a height of 500 feet above the river.

CONCORD & MONTREAL RAILROAD. The principal connecting link, with the Southern Division of the Boston & Maine Railroad, between Boston and the mountain region of New Hampshire. Through its branches it brings even the lofty summit of Mt. Washington within a comparatively few hours' ride of New York and Philadelphia. It traverses the beautiful valley of the Merrimack, following the river as far as Concord, N. H., near which it is crossed for the last time. The road then leaves the river banks, and, passing through Tilton, Laconia, and Lake Village, and a region unsurpassed for the picturesqueness of its scenery, enters upon the banks of Lake Winnipesaukee. At Laconia connection is made with the Lake Shore Branch,

which skirts the southerly shores of the lake to
Alton Bay. At The Weirs, where connection is
made by the main line by steamboat with Centre
Harbor, Wolfeboro, and other points on the lake,
this beautiful body of water is seen stretching out
to the eastward, while the distant mountains form
a blue border of exquisite loveliness. [See Lake
Winnipesaukee.] At Plymouth connection is made
with trains on the Pemigewasset Valley Branch,
which runs to North Woodstock, and leaving the
Pemigewasset River, the main road ascends the
valley of Baker's River, passes through Rumney,
with Mt. Stinson upon the right, and the Mt. Carr
Range upon the left. Farther along Rattlenake
Mountain is seen on the right, and high hills are
on every side. The highest point on the line, 1063
feet above the sea, is reached at Warren Summit.
Mts. Carr, Waternomee, and Kinneo are upon the
right, and Mt. Mist and Webster's Slide upon the
left. At East Haverhill two prominent mountains
on the right are Black and Sugar Loaf. At Haverhill, a few miles further on, the Connecticut valley
is reached, and on the opposite side of the Connecticut River is Mt. Pulaski, and in the distance
down the river is to be seen the shapely peak of
Mt. Ascutney. The road now follows the banks
of the Ammonoosuc River, which enters the Connecticut at Woodsville, passing through a succession of the most picturesque scenery to Littleton.
At Wing Read the railroad branches off to Bethlehem Junction, where connection is made with the
Profile & Franconia Notch Railroad, which runs
down to the Notch, while the main road continues
on past the Twin Mountains to Fabyans. From
Wing Road the principal line connects at Whitefield with the Whitefield & Jefferson Branch, taking passengers to all of the principal resorts north
of the Presidential Range, and to Lancaster and
Groveton, where it connects with the Grand Trunk
Railway.

CONWAY, N. H., on the Northern Division of the
Boston & Maine Railroad, a village locally known
as "Conway Corners," and anciently bearing the
name of "Chatauque," is pleasantly situated about
five miles south of North Conway [see North
Conway], in the valley of the Saco, at the conflu-
ence of the Saco and Swift rivers, and amidst
peaceful and rural scenery. It is a favorite re-

treat of persons preferring quiet life to the bustle
of a crowded resort. All the picturesque places
within easy distance from North Conway can be
readily reached, while the village is half a dozen
miles nearer to such popular objects of interest as
Chocorua Lake and Mountain, Jockey Cap, Moat
Mountain, Ridge Road and Lovell's Pond.

COPPER MINE BROOK. See Franconia.

COPPLE CROWN. A small mountain five miles from
Wolfboro. [See Wolfboro.] It is but about 2100
feet high, and may be easily ascended.. Carriages
can be used, if desired, to convey the tourist from
the village to within a mile of the summit, and the
ascent can be made on horseback. The view from
the summit is very fine. Lake Winnipesaukee,
which forms a part of all the views in this vicinity,
is visible for nearly its whole length. Belknap
and Gunstock, with the mountains of the Merrimack valley, stretch away toward the west. To
the south is a comprehensive view of forest and
meadow, with ponds and villages dotting the landscape. The Ossipee and Sandwich ranges tower
above the lake to the northwest. Almost due
north, Chocorua, with Mount Washington high
above it, indicates the White Mountain region.
The ocean can be seen on a perfectly clear day.

CRAWFORD, ABEL, the "Patriarch of the Mountains," for whom the Crawford Notch was named
[see Crawford Notch]. In its midst he lived and
reared his family, and at the age of 70 he made the
first ascent ever made to the summit of Mt. Washington on horseback. When he was 80 years of
age he was accustomed to walk to his son's house
at the "Gate of the Notch" before breakfast, a
distance of five miles. He was one of the first
guides of that region.

CRAWFORD BRIDLE PATH, before the Mt. Washington Carriage Road and Railroad, the principal
route to the summit of Mt. Washington. Bayard
Taylor calls it "by far the most compensating
road to the summit," and its great advantage consists in the fact that it affords a view from several
mountain peaks. It was cut by Ethan Crawford
[see Ethan Crawford], and is now disused for
horses, but pedestrians find it an interesting and
delightful feature of mountain travel. It is about
nine miles long and in places very steep and rough.
It starts from the base of Mt Clinton, in Crawford

Plateau [see Crawford Plateau], and ascends for two or three miles through a dense forest to near the summit, 4320 above the sea. At this height the trees have dwindled away to a few stunted firs, growing from between the crevices of the rocks. Just before reaching the top, a region of dead trees, supposed to have been killed by the intense frosts of 1812 and 1816, are passed. The path lies a little to the north of the summit, giving a wonderful and comprehensive view. Towards the east, almost directly in front, is to be seen the conical summit of Mt. Kearsarge, and behind, Mt. Willard and the other mountains around the Notch. In descending the narrow ridge which joins Mt. Clinton, on the right is to be seen, at the depth of 2,000 feet, the vast forest through which win is the Mt. Washington River, and beyond is a long range of hills, and on the left at a similar depth the Ammonoosic River. The path passes around the southern side of Mt. Pleasant, several hundred feet below the summit, and enters on a plain lying at the foot of Mt. Franklin, and extends to near its summit. The ascent here is quite arduous, the path winding along to the northwest of the mountain. By leaving the path for a short distance to the right, the highest point may be reached, and a grand view to the southeast obtained. Far to the south may be seen the four beautiful peaks of Chocorua. On the almost perpendicular eastern side of the mountain, can be seen after it is passed, the long scar left by a slide which occurred in the summer of 1857. The path now passes around the southeastern side of Mt-Monroe, several hundred feet below the summit, on the right of which is Oake's Gulf. [See Oake's Gulf.] From this point the first view of Mt. Washington is obtained, the summit, which looks like an irregular pyramidal pile of great brown stones, is nearly fifteen hundred feet above. The cone of the mountain is reached by way of an extended plateau, which is at first quite smooth, but nearer the mountain is covered with great boulders. The path winds among these rocks, and is marked in various ways. The ascent of Mt. Washington is made from the southwestern side, and is not difficult.

CRAWFORD, ETHAN ALLEN, the "Giant of the Hills," son of Abel, cut the first bridle path [see

Crawford Bridle Path] to the summit of Mt. Washington in 1821. He resided at that time near the Giant's Grave on the north side of the Crawford Notch. [See Crawford Notch.] In fact, all of the old paths on the western side of the mountains were cut by the Crawfords. Ethan was famous not only as a guide, but as a hunter.

CRAWFORD OR WHITE MOUNTAIN NOTCH. A narrow pass, about 12 miles long, lying between Mts. Willard, Willey, and Nancy on the west, and Webster, Jackson, and Crawford on the east. There is probably nothing grander in the way of scenery than is encompassed within those few miles, and the view obtained from the railroad track winding along the sides of the mountains, is even finer than that obtained from the carriage road. In going through the Notch on the Maine Central from North Conway, the tourist follows up the course of the Saco River and Rocky Branch River, and enters the Notch at Bartlett, a small village, where observation cars are provided. A little to the left are the three peaks of Tripyramid Mountain, about 4200 feet high. A high rock close to the carriage road on the left hand is called Sawyer's Rock [see below], and a little beyond this is Hart's Ledge and a small stream called Sawyer's River. The first point of special interest is Nancy's Brook, named for a girl from Jefferson, who, in the winter of 1788, followed her runaway lover through the wilderness of the Notch, and, becoming exhausted, was frozen to death near the brook. Half a mile beyond is Bemis Station where is to be seen the old Mount Crawford House, one of the oldest and, in its day, most popular houses of the region, kept by old Abel Crawford. [See Abel Crawford.] It is now a farm house. Up the valley of the Saco is to be seen Mt. Crawford, the southern peak of which is 3134 feet high, and the northern, 3500. Between them is Mt. Resolution, 3400 feet high. Looking up the gorge, Mt. Webster, 4000 feet high, is to be seen on the right, and Mt. Willey, 4300 feet high, on the left. The rounded summit of Mt. Willard, 2570 feet, forms a background to the picture, and the Giant's Stairs are distinctly visible beyond the river on the right. The train now passes over that wonderful piece of engineering skill, the Frankenstein trestle, 500 feet long and 80 feet high, and built entirely of iron.

Two miles further on the famous Willey House is seen far down in the valley, the scene of the terrible tragedy enacted by the elements in 1826, when Samuel Willey, jr., his wife and five children, and two hired men perished by being buried in a landslide that occurred on the 28th of August of that year. The bodies of three of the children still lie engulfed in the mass of rocks and earth that, precipitated from the side of Mt. Willey, covered a space in the valley nearly a mile in length. The railroad winds around the side of the mountains several hundred feet above the Saco for a number of miles, affording excellent views of the scenery of the valley. Willey Brook is crossed on a bridge nearly 100 feet high, and the train continues along the foot of Mt. Willard, passing through the northern gate of the Notch, a narrow cut 50 feet deep in the solid rock, and stopping in front of the Crawford House, a short distance beyond. From here there is a bridle path to the summit of Mt. Washington, long disused except by pedestrians. Mt. Willard is easily ascended by pedestrians from here, and the view from its summit, especially in the afternoon, has attractions to be found only at this point. Near the summit is a remarkable cavern known as the "Devil's Den," which, however, is only accessible by means of ropes. Among the places of interest in this vicinity are the "Old Maid of the Mountain," a great stone face on a spur of Mt. Webster, the "Devil's Pulpit," near the Gate of the Notch, a profile rock called "The Infant," opposite the "Devil's Pulpit," another profile, the "Young Man of the Mountain," Elephant's Head, the Flume, the Silver Cascade, the Sparkling Cascade, the Sylvan Glade Cataract, and Gibbs's Falls, all possessing remarkable charms distinctively their own. [See these places under their appropriate titles.] Continuing the journey through the valley for four miles, Fabyan's is reached. The great White Mountain Notch was unknown to the white inhabitants until about the year 1771, when a hunter named Nash, if tradition is to be believed, in climbing a tree on Cherry Mountain to look for game, perceived what he thought to be an opening through the unexplored mountains. Making his way in a southeasterly direction, he arrived at the point now known as the Gate of the Notch, then a mere gorge, through which he entered the great

passageway between northern and southern New
Hampshire. Proceeding to Portsmouth, he an-
nounced the welcome discovery to Governor Wentworth, who, to test the feasibility of the pass,
promised Nash a large tract of land on the north
side of the mountains if he would get a horse
through and bring him to Portsmouth. Enlisting
the aid of a fellow-hunter named Sawyer, he suc-
ceeded in the undertaking, though having sometimes to draw the horse up high precipices with
ropes, and let him down on the other side in the
same manner. It is told that when they had safely
deposited the horse in this way on the last rock
on the south side, the jubilant and bibulous Sawyer drained the last drop from a flask of rum, and,
dashing the bottle on the rock, exclaimed, "This
shall hereafter be called Sawyer's Rock!" and it
is known by that name to this day. The land re-
ceived as a reward for this feat is now known as
Nash and Sawyer's Location.

CRAWFORD NOTCH CARRIAGE ROAD, the tenth
turnpike built in New Hampshire, was incorpora-
ted in 1803. It extends through Crawford Notch
[see Crawford Notch], and Nash and Sawyer's location, 20 miles, and cost $40,000. Before it was
laid out, the old county road crossed the Saco
River 32 times in making its way up the valley.
Since the railroad was built through the Notch
there has been little travel over the carriage road
except in summer.

CRAWFORD PLATEAU, a station on the Maine Central Railroad, situated at the northern extremity
of Crawford Notch [see Crawford Notch], 2000 feet
above the sea level. It is the highest point in
the valley, and water flows from it in both directions. Near the "Gate of the Notch" is the site
of the old Notch House, erected by Ethan Allen
Crawford and his father, and kept for years as a
public house by Thomas, a brother of Ethan. It
was for a long time the largest house of the region.
Here the Saco River has its source in a pond called
Saco Lake. Numerous paths lead to places of interest near by. There is a carriage road to the
summit of Mt. Willard [see Mt. Willard], and a
bridle path to the top of Mt. Washington. [See
Crawford Bridle Path.]

CRYSTAL CASCADE, a beautiful waterfall in Pinkham Notch [see Pinkham Notch], three miles from

the Glen [see The Glen], which is partially fed
from the dome of Mt. Washington. It is reached
by a pathway through the woods about a third of
a mile in length, marked by a guideboard at the
entrance from the carriage road. A rustic bridge
crosses the stream below the fall from which an
excellent view is obtained. "Some 70 feet above,"
quoting Eastman's White Mountain Guide, "we
can see the brook pouring in a single stream
around the bend. Then the rock broadens into a
rough stairway, with easy slope, which grows
wider and wider to the bottom, and down these
steps the spreading water sheds its white, thin,
dancing, and broken sheet, showing, now and then,
through its gauzy texture, the deep green mosses
clinging to the rocks, which soften its own fall."

CUSHING ISLAND, ME., at the mouth of Portland
Harbor, and four miles from the city, is bold and
rocky, its surface gradually rising from the harbor
side to its southeastern shore, which, at a consid-
erable altitude, presents a precipitous front to the
sea, terminating at the northeastern end in a cas-
tellated bluff of perpendicular rock nearly 150 feet
high. The island is 250 acres in extent, and is covered with a dark forest growth along a high ridge
for its entire length. From these woods the land
descends on the harbor side to an arable valley,
and thence to the beaches and ledges which line
the inner shore. From White Head, a lofty precipice, a fine view is obtained of the ocean, the bay,
and the city. Good beaches afford opportunities
for sea bathing, while both the shore and deep-sea
fishing are excellent. Fine roads make it possible
to drive for six or eight miles along the shore
among beautiful summer residences. A steamboat runs between the island and Portland in the
summer.

D

DAVIS BRIDLE PATH. A road to the summit of
Mt. Washington, starting from near the old Mt.
Crawford House in the Crawford Notch [see Crawford Notch], which is longer than the other paths,
but inferior to none in romantic interest.

DENNING'S LAKE. See Southwest Harbor.

DEVIL'S DEN. A suggestive name given to many
of the small caves in the mountain regions of New
England. The most remarkable cavern bearing

this name is on the southern side of Mt. Willard
[see Mt. Willard], near the summit, the mouth of
which can be distinctly seen in coming up Crawford Notch. It is only accessible by means of
ropes from above, and is about 20 feet wide, 15
feet high, and 20 feet deep. It is so cold and damp
that neither birds nor beasts inhabit it.

DEVIL'S PULPIT. See Crawford Notch.

DIXVILLE NOTCH, characterized by Dr. Jackson,
the geologist, as more Alpine in its character than
any other mountain pass in New England, is in
the extreme northern portion of New Hampshire,
about 60 miles beyond the White Mountain Range.
The pass is much narrower than either the Crawford or Franconia notches, and is much shorter than
they are. Its length is but a mile and a quarter,
but it has more of the character of a notch than
the two mentioned. So narrow is the pass that
the roadway could only be constructed by building
up against the mountain's side a substructure of
rude masonry, not quite wide enough to accommo-
date two carriages abreast except at certain points
where turn-outs are provided. The walls of the
mountains slope upwards so sharply on either side
that much labor and expense is required every
year to clear the road of the stones and earth that
are released from the sides of the mountains by
the frosts and rain. The decaying cliffs of mica
slate which overhang the way shoot up in most
singular and fantastic shapes, and vary in height
from 400 to 800 feet. It is believed that some have
crumbled away to half their original height. The
whole aspect is one of wreck and ruin. Table Rock,
much the highest pinnacle, juts out from the southerly wall of the pass about 100 feet above the road.
A very steep and difficult path leads to it. It is
only some six or eight feet wide and about 150 feet
long, forming the top of the pinnacle, overhanging
an almost unbroken precipice on each side of several hundred feet. From it Maine, Vermont, and
Canada can be seen. A few miles to the east is
Lake Umbagog, and about 10 miles to the north is
Lake Connecticut, the source of the Connecticut
River. A path leads from Table Rock to Snow
Cave, near by, a deep wedge-shaped crevice in the
mountain, in which ice remains until late in August. The profile of a man can be discerned in the
face of this cliff, as seen from below. Just before

the eastern gateway of the Notch is reached, is to be seen a flume, and on the opposite side of the road in the woods, just beyond the Notch, there is a series of beautiful cascades extending nearly a mile. The Notch is reached by train from Boston over the Concord & Montreal Railroad to Lancaster, N. H., and thence by stage to Colebrook, N. H. The Notch is about 10 miles from there.

DOG MOUNTAIN. See Southwest Harbor.

DOUBLEHEAD MOUNTAIN. See Jackson.

E

EAGLE MOUNTAIN. See Some's Sound.

EAST CHOP LIGHT. See Martha's Vineyard.

EAST TILTON, N. H. A station on the Concord & Montreal Railroad, 97 miles from Boston, remarkable for giving the tourist over this route the first fine view of the distant mountains. As the train rounds Sandbornton Bay, and approaches the station, the Sandwich Range is seen. The peak on the left is Sandwich Dome, and in order, toward the right, are Tripyramid, Whiteface, Passaconoway, Chocorua, and the Ossipee Range. Across Lake Winnisquam are to be seen the distant peaks of Moosilauke, Lafayette, Kinneo, Cushman, Liberty, Tecumseh, and other mountains of the Franconia Range.

ECHO LAKE. A small and beautiful sheet of water a short distance to the north of the northerly gate of the Franconia Notch. [See Franconia Notch.] It is remarkable for being the centre of marvellous echoes.

EDGARTOWN. See Martha's Vineyard.

ELEPHANT'S HEAD. See Crawford Notch.

ELEPHANT ROCK. See Newport, N. H.

ELLIS RIVER. See Pinkham Notch.

ENDICOTT ROCK. A curiously and quaintly carved stone near The Weirs [see The Weirs], in midchannel of the lake's outlet. It was accidentally discovered a number of years ago, and is supposed to be a monument or boundary mark made by two surveyors sent out by Gov. John Endicott of Massachusetts. It is inscribed with Gov. Endicott's name, and the initials of Edward Johnson and Simon Willard, who were commissioned to find the head of the Merrimack. The rock has been raised from its bed in the stream at the point where it flows from the lake, and the greater part of it is above the surface of the water. It may be seen on the right just before the train going north reaches The Weirs station.

ENFIELD, N. H. A village situated in the midst of scenery of surpassing loveliness. It is 133 miles from Boston, on the Concord Division of the Boston & Maine Railroad. There are several very fine ponds in the vicinity, besides Mascomo Lake [see Mascomo Lake], on the eastern side of which is one of the most prosperous Shaker communities in the country.

ETHAN'S POND. A small body of water on the summit of Mt. Willey, which, although but a short distance from the Saco River, finds its way towards the southwest, and empties into the Pemigewasset River.

F

FABYAN'S. One of the most noted and popular resorts, is situated in the very heart of the White Mountain region at the junction of the Concord & Montreal and the Boston & Maine railroads, 208 miles from Boston. A noble view of the Presidential Range is to be obtained from here, and the lengthened scars on the side of Mt. Pleasant, a thousand feet in height, and said to represent an Indian chief with tomahawk, are plainly visible. Near here are the once famous Lower Ammonoosuc Falls, which have been spoiled by the erection of a sawmill above. The river is said to be the wildest and most rapid of all the New Hampshire rivers, falling nearly 6000 feet in its course of 30 miles from Mt. Washington to the Connecticut River. It has many cascades, one of which, the Upper Ammonoosuc Falls, are three and a half miles from Fabyan's on the carriage road to Mt. Washington.

FERNALD'S POINT. See Some's Sound.

FERRIN'S POND. See Profile Lake.

FIFTEEN MILES FALLS. See Littleton.

FLUME, THE. A great fissure in the rocks at the base of Flume Mountain at the southerly end of Franconia Notch, and easily reached by a good carriage road from the main road running in a southeasterly direction about a mile. The Flume is about 700 feet long and from 60 to 70 feet in depth. The width between these perpendicular walls of granite is a general average of 20 feet, except at the upper end, where they suddenly contract to about 10 feet, and formerly held suspended between them, about midway up their sides, a huge bowlder of granite, which at some remote period must have come crashing down the mountain sides till its further descent was interrupted at this wonderful place. On June 20, 1883, a great storm occurred in the mountains, causing several terrific land slides, one of which, starting nearly at the top of Mt. Liberty, gathered force when it reached the Flume stream, and swept down the narrow defile with resistless fury and tremendous force, carrying rocks weighing many hundred tons, and extending the high walls of the Flume some 500 feet. The most serious damage done was the displacement of the great bowlder, which fell with a terrific crash into the ravine below. It was subsequently found lodged some distance below the mouth of the Flume. Its presence had added greatly to the wildness of the scene, and so nicely was it adjusted, and so slight appeared its hold, that it gave the impression to one standing under it that the slightest touch would be sufficient to send it crashing from its resting place into the ravine below. At the upper end of the Flume a new cascade was formed by the torrent, and an immense rock was thrown over upon others, while the under side was hollowed out so as to leave a grotto of respectable size. A plank walk extends up through the Flume beside, and at times partly over the stream, which comes tumbling tumultuously along its rocky bed, and one can follow it through this narrow gorge, and, by climbing the rocky heights, can obtain an excellent view from above. A bridge across the chasm has been formed by the fallen trunk of a tree. There is a path from the Flume to the Pool. [See Pool, The; also, Franconia Notch.]

FLUME CASCADE. One of a series of beautiful falls, not excelled for beauty in the whole range of mountain travel, a short distance from Crawford Plateau. [See Crawford Plateau.] From this point a full view of the summit of Mt. Washington is to be had. A path up the side of the mountain for about a third of a mile leads to the Cascade.

FLYING MOUNTAIN. See Southwest Harbor.

FRANCONIA. A town in New Hampshire situated

just north of Franconia Notch [see Franconia
Notch], and in the valley of Gale River, six miles
over a pleasant stage road from Littleton, and a
little less from Bethlehem. The approach from
the latter place is over a spur of Mt. Agassiz, and
the descent into the deep, bowl-shaped valley sup-
plies one of the most delightful drives in the
mountains. The nearness of Mt. Lafayette and
Mt. Garfield and the other grand peaks of the
Franconia Range gives to the view a startling
boldness, while the foreground of meadow and
forest adds rare picturesqueness and beauty. Gale
River and Lafayette Brook afford excellent trout
fishing, and other mountain streams are also accessible. In the vicinity are Bridal Veil Falls,
Copper Mine Brook, and Mt. Kinsman Flume.
[See Bridal Veil Falls.] In the village are the
buildings of the Franconia Iron Company, which
began mining operations here in 1805.

FRANCONIA NOTCH. A narrow pass about five
miles long lying between Mts. Pemigewasset,
Kinsman and Cannon on the west, and Flume,
Liberty, Lincoln, Lafayette and Eagle Cliff on the
east. From the Flume House, at the foot of Pemigewasset Mountain, to the little plateau on which
the Profile House is situated, at the north end of
the Notch, there is an ascent of 543 feet, and in
places the pass is very narrow. An excellent carriage road winds through the forest, with frequent
glimpses of the high beetling cliffs on either side.
The Pemigewasset River, here a tumbling, rapid
stream, is crossed a little distance above the Flume
House, from which there is a path through the
woods to Georgianna Falls [see Georgianna Falls],
two miles below. About three quarters of a mile
from the hotel, in a southeasterly direction, are
the Cascade and Flume [see Cascade and Flume],
and about the same distance in a more easterly
direction is the Pool. [See The Pool.] Continuing
up the road for a mile and the Basin [see The Basin]
is reached. A mile beyond a mountain brook crosses
the road, and a walk of half a mile up its banks,
brings one to Walker's Falls. [See Walker's Falls.]
The frowning southern cliffs of Cannon Mountain
are seen in front for some distance, while Eagle
Cliff rises as the eastern wall of the Notch. A
half mile further up the road is a clearing where
once stood the Lafayette House, which was de-

stroyed by fire in the spring of 1861, and a mile and
a half beyond is the trout house where fish are
bred. A short distance beyond is Profile Lake
[see Profile Lake], and when it is nearly passed
the rugged features of the famous profile itself
come plainly into view. [See Old Man of the
Mountain.] A half mile further on is a bridle
path to the summit of Mt. Lafayette, and near by
a disused carriage road leading to the summit of
Ball Mountain. A short distance to the north is
Echo Lake. [See Echo Lake.] Here ends the Notch.
The Profile & Franconia Notch Railroad, a narrow-gauge road of peculiar construction, enters the
Notch at this point and has its terminus at the
hotel. It extends from here to Bethlehem Junction, a distance of nine miles, where it connects
with trains on the Concord & Montreal Railroad,
and with a branch road four miles long leading to
Bethlehem village.

FRANKENSTEIN TRESTLE.   See Crawford Notch.

G

GALE RIVER.   See Franconia.

GARNET POOLS.   See Gorham.

GAYHEAD.   See Martha's Vineyard.

GEORGIANNA FALLS. A beautiful cascade on
Harvard Brook, between one and two miles from
North Woodstock, N. H. [See North Woodstock.]
It can also be reached by a path through the
woods from the Flume House about two miles
long. This place was visited for the first time in
September, 1858, by a party of gentlemen who,
with the ceremonies usual on such occasions, gave
the cascade its name. The honor of discovery is
also claimed by some Harvard students, and the
brook still retains the name of their Alma Mater.
It is the largest waterfall of any previously discovered among these mountains, the water dashing
down a precipice, through a mountain gorge,
nearly one hundred and fifty feet. For about a
mile there is a series of cascades of indescribable
beauty. There is also to be seen a charming little
sheet of water in one of the hollows of the rock,
which has been named "The Mirror," from its
wonderful transparency.

GIANT'S STAIRS, THE.   See Crawford Notch.

GIBBS'S FALLS, A romantic cataract about half an

hour's walk from Crawford Plateau [see Crawford Plateau]. The water of a brook makes an abrupt descent of 30 or 40 feet in two distinct sheets, which are separated by a projecting cliff.

GLEN, THE, a plateau 830 feet above Gorham, N. H., valley, and 1632 feet above tide water at Portland. It is a magnificent mountain bowl, formed by Mts. Washington, Clay, Adams, Madison, Jefferson, Carter and Wild Cat There is a carriage road to Gorham [see Gorham], eight miles, and to Jackson [see Jackson], through Pinkham Notch [see Pinkham Notch], 12 miles. The distance to Glen Station [see Glen Station], is 16 miles, via Jackson. The most noted carriage road, however, is the famous Mt. Washington Carriage Road [see Mt. Washington Carriage Road] which has its terminus at the Glen. The view of the mountains from this point is unsurpassed, and many places of interest are in the immediate vicinity, including Osgood's Falls and Raymond's Cataract.

GLEN ELLIS FALL, a cataract of exceeding beauty, in Pinkham Notch [see Pinkham Notch], and about four miles from the Glen [See The Glen.] An excellent plank path from the road — marked by a guide-board — into the forest leads to the great wall of rock about a quarter of a mile from the rock over which the Ellis River leaps into the pool 100 feet below. The stream first slides through a deep groove it has worn in the granite at a very sharp angle for some 20 feet, and then leaps as from the nose of a great pitcher 60 feet more. On the opposite side is the steep wall of Wild Cat Mountain, more than 3000 feet high, and there is probably not a wilder or more romantic spot in the mountain region than this. A series of wooden stairways lead to the depths below, where the green, placid surface of the basin forms a striking congrast to the wild, leaping torrent that feeds it.

GLOUCESTER, MASS., 28 miles from Boston, an old fishing town, reached by steamer and the Eastern Division of the Boston & Maine Railroad, is surrounded by fine points of view seaward, beaches, and rocky cliffs. Close to the town is Eastern Point, a rocky promontory, on the peak of which are the remains of an old fort; and directly across the harbor is the rugged and picturesque East Gloucester Peninsula. Excursions can be made to "the reef of Norman's Woe," where occurred

"The Wreck of the Hesperus," immortalized by Longfellow. It is about two miles from Gloucester, and is a sombre and threatening mass of rocks. About a mile southwest of Norman's Woe is Rafe's Chasm, an enormous fissure, 10 feet wide, 60 feet deep, and 100 feet long, which yawns into the cliffside. Not far off is another wonderful fissure in the trap-rock; and beyond is Goldsmith's Point, with its colony of summer villas.

GOLDSMITH'S POINT. See Gloucester, Mass.

GOOD HARBOR BEACH. See Bass Rocks.

GOODRICH FALLS, the largest perpendicular fall among the White Mountains, is about a mile below Jackson, N. H. [see Jackson], on the Ellis River near its junction with the Wild Cat River. It is but a short distance on the right from the carriage road to Glen Station. After heavy rains the view of the fall is very grand, and at all times the scenery is wild and picturesque.

GORHAM, N. H. A thriving village at the junction of the Peabody and Androscoggin rivers, 880 feet above the sea, and a station of the Grand Trunk Railroad. It is the only point from which the beauty of the range of Moriah, Carter and the Imp can be seen to advantage. Mt. Carter is about 5000 feet high, and Mt. Moriah is some 200 feet lower. Between the two is the Imp, so named from the marked resemblance the summit bears to a grotesque human countenance when viewed from a particular point. Paths lead to the summits of all these mountains. To the northwest is the Pilot range of hills, and to the east and southeast the Androscoggin hills. The summit of Mt. Madison is visible from Gorham. Randolph Hill, from the summit of which one of the best views of the Mt. Washington Range can be obtained, can be reached by a carriage drive of about five miles. The whole northerly wall of the range, from valley to crest, is seen to the very best advantage. Berlin Falls [see Berlin Falls], on the Androscoggin, are six miles distant. Mt. Surprise, a spur of Mt. Moriah, is also accessible from Gorham, as is Mt. Hayes. [See Mts. Carter, Moriah, Surprise and Hayes.] A pedestrian excursion to the summit of Mt. Madison across the northerly ridges of the Presidential Range can be made from the foot of Randolph Hill with the assistance of a guide. The course is over Mt. Madison, around or over the sharp pyramid of Adams, over Jefferson, between the humps of Mt. Clay to the top of Mt. Washington, nearly a day's tramp. This route brings into view all the great ravines of the range, excepting Tuckerman's, the long and narrow gulley between Mts. Adams and Madison, King's Ravine, named for Starr King, the gulf between Mts. Adams and Jefferson on the southeast, and the gorge in Jefferson on the northwest. There is a good carriage road from Gorham over what is known as the Cherry Mountain Road to Jefferson, thence over the mountain to the Crawford Notch, or to Lancaster, or to the Franconia Notch. [See Cherry Mountain, Jefferson, Lancaster, Franconia Notch.] A series of basins in the Peabody River, near the Gorham road, curiously hollowed out of the rocks and delicately polished by the water, have been named Garnet Pools. A drive of eight miles over a road that, for the greater part of the way, follows the easterly bank of the Peabody River, brings one to the Glen and Pinkham Notch, through Jackson to Glen Station, and through the Crawford Notch. [See Pinkham Notch, Jackson, Glen Station, and Crawford Notch.]

GRAFTON NOTCH (also called Bear River Notch), a mountain pass in Maine, in which there is a remarkable flume known as Moose Chasm. Its the walls are as high and perpendicular as those of Flume [see The Flume], and between flows a large and noisy stream which plunges beneath a mass of superincumbent rock, and disappears from sight. Grafton Notch is on the road between Dixville Notch and Bethel, Me. [See Bethel.]

GREAT GULF. See Gulf of Mexico.

GREAT HEAD. See Bar Harbor.

GREAT SQUAM. See Asquam Lakes.

GREEN MOUNTAIN. See Mt. Desert.

GREEN MOUNTAINS. The northern portion of the great Appalachian chain which for many hundred miles fronts the Atlantic coast of the United States. Their wooded sides obtained for them from the early French explorers the name of Monts Verts, and from these words is derived the name of the state (Vermont) in which they are situated. Without attaining any great elevation, these hills form an unbroken watershed between the affluents of the Connecticut River on the east side of the Hudson River and Lake Champlain on

the west, and about equidistant between them.
South from Montpelier, two ranges extend, one
toward the northeast, nearly parallel with the
Connecticut River, dividing the waters flowing
east from those flowing west; and the other, which
is the higher and more broken, extending nearly
north, and nearer Lake Champlain Among the
principal peaks are Mt. Mansfield, 4348 feet above
the sea; Camel's Hump, 4188 feet; Killington Peak,
3924 feet; and Ascutney, 3320 feet.

GULF OF MEXICO. An immense amphitheatre or
ravine between Mts. Washington and Clay, down
which one can look for nearly a thousand feet
from the Mt. Washington Carriage Road or Railroad. It is also called the Great Gulf.

H

HAMPTON BEACH, N. H., reached by stages from
Hampton, 46 miles from Boston, is a much frequented resort. Boar's Head is a lofty headland
extending into the sea and dividing Hampton
Beach from Rye Beach. [See Rye Beach.] The
view from its top is very grand. The bathing and
fishing here are excellent, the scenery fine, and the
drives pleasant.

HARVARD BROOK. See Georgianna Falls.

HAVERHILL, N. H. A charming village on the
Concord & Montreal Railroad, 160 miles from Boston. It is situated on the east bank of the Connecticut River, and Newbury, Vt., is on the opposite side beneath the heights of Mt. Pulaski. Mt.
Ascutney, a shapely peak, is to be seen in the distance down the river.

HERMIT LAKE. See Tuckerman's Ravine.

HIGHLAND LAKE. See Bridgton, Me.

I

ICE GORGE. See Randolph.

IMP, THE. See Gorham.

INDIAN ROCK. See Rangeley Lakes.

INFANT, THE. See Crawford Notch.

ISLES OF SHOALS, N. H., a group of nine bare and
rugged islands, lying about nine miles off the
coast, and reached from Boston and Portsmouth in
the summer by daily lines of steamers. The
islands are small in extent, the largest — Apple-

dore — containing only 350 acres. They all have a
bleak and barren aspect, with little vegetation,
and with jagged reefs running far out in all directions among the waves. Appledore rises in the
shape of a hog's back, and is the least irregular in
appearance. Its ledges rise some 75 feet above
the sea, and it is divided by a narrow, picturesque
little valley. Just by Appledore is Smutty Nose
or Haley's Island, low and flat, with threatening
reefs. On it are to be seen the graves of fifteen
sailors, rudely marked, who formed the crew of
a Spanish ship wrecked there in 1685. There is
also a little graveyard containing the remains of
the hardy pioneer who built a sea wall that yet remains a monument to his humanity and industry.
The story of this achievement is told on the tomb-
stone that marks his grave, as follows: "In memory of Mr. Samuel Haley, who died Feb. 7, 1811,
aged 84. He was a man of great ingenuity, industry, honor and insight; true to his country, and a
man who did a great public good in building a
dock and receiving into his inclosure many a poor,
distressed seaman and fisherman in distress of
weather." About a quarter of a mile beyond is
Star Island, formerly the site of the odd little village of Gosport. A place of interest is the stone
church which crowns the crest of the highest
point. It has a wooden tower on one end, with a
good-toned bell, which is always rung in foggy
weather, of which they have but little. Over the
door is the legend, "Gosport Church — originally
constructed of the timbers from the wreck of a
Spanish ship, A. D. 1685. Was rebuilt in 1720, and
burned by Islanders in 1790. This building of
stone was erected A. D. 1800." It is about 30 feet
long. Its walls are a foot and a half thick. It has
20 stiff-backed pews, which will hold five persons
each. The pulpit is small and not particularly or-
namental. Behind it are two stiff, old-fashioned
chairs, with chintz-covered cushions. During the
vacation season services are held therein. Another relic of the olden time is Fort Star, formed
of stone and earth, about 60 feet square. It was
erected about 1653. "to withstand foreign enemies
and to protect the commerce of this island. Restored in 1692 and garrisoned by a company of Pro-
vincial forces during the bloody French and Indian
war. In 1745 this fort was again repaired and

manned with nine guns. Dismantled in the Revolution and the guns sent to Newburyport." On the west toward the mainland is Luncheon Island, jagged and shapeless, with a diminutive beach, while two miles away is the most dangerous and forbidding of all these islands, Duck Island, many of the ledges of which are hidden insidiously beneath the water at high tide, and at low tide are literally covered with seagulls. White Island, the most picturesque of the group, is about a mile southwest of Star Island, and has a powerful revolving light, visible for fifteen miles around. The other three islands are Cedar, Mallagar and Seavy.

J

JACKSON, N. H., a small village situated on the stage route between North Conway and the Glen, three miles from Glen Station [see these places], in the midst of grand mountains on the Wild Cat River, a tributary of the Ellis River, which it enters a short distance below. Jackson Falls, on the former, and Goodrich Falls [see Goodrich Falls] on the latter, are places of interest. Fine views are to be obtained of Iron Mountain, 2900 feet high, the bald peak of Tin Mountain, the two peaks of Double Head, one 3000 and the other 3100 feet high, Thorn Mountain, Wild Cat Mountain and, from a point near by, the house on the summit of Mt. Washington. From Jackson excursions can be made to the summit of Mt. Washington, Glen Ellis Falls, Crystal Cascade, Tuckerman's Ravine, North Conway, Crawford Notch, Pinkham Notch, Carter Notch, Thorn Mountain, Winnewetah and Appalachian Cascades [see these titles], and other less notable places. Jackson abounds in beautiful walks and drives.

JAY PEAK. See Newport, Vt.

JEFFERSON, N. H. A village on the Whitefield & Jefferson Branch of the Concord & Montreal Railroad, 210 miles from Boston. It is in some respects a rival of Bethlehem. It is similarly situated upon an elevation, and the pure air and general healthfulness of the locality has made it a great resort for invalids and pleasure-seekers. The outlook upon the Presidential Range, with Mts. Adams and Jefferson in the near front, is extremely grand. Starr King declares "Jefferson

Hill may, without exaggeration, be called the ultima Thule of grandeur in an artist's pilgrimage among the New Hampshire mountains; for at no other point can he see the White Hills themselves in such array and force." In fact, every one of the great White Mountain group is visible, and the railway and hotel on Mt. Washington can be distinctly seen. There is also plainly outlined the Franconia Mountains, the side of the Willey Mountain in the Notch, and the line of the nearer Green Mountains beyond the Connecticut,—in fact, a panorama of hills to the northwest and north, almost as fine as the prospect in that direction from the summit of Mt. Washington. Mt. Pliny is the ancient name for the long wooded range in the northeast part of the town; and in 1861 the culminating part of the range was named Mt. Starr King in honor of the author of "The White Hills." It is easily ascended, and affords one of the best views to be had of the Presidential Range. The memory of the celebrated divine is warmly cherished by the people of Jefferson, and they fondly recall the hours he used to spend lying on the grass and rapturously describing the charms of this his favorite side of the mountains. It was also, prior to the war, a great resort for Southerners, and the landlords never tire of telling of the time when they numbered among their guests Calhoun, Randolph and other Southern statesmen. There are two ponds in the town,— Cherry Pond, upon the line of the railroad, and Pond of Safety. [See Pond of Safety.] The former is the chief source of John's River and the latter of the Upper Ammonoosuc. Israel's River traverses Jefferson from the southeast to the northwest. The two rivers, John's and Israel's, derived their names from two brothers, John and Israel Glines, who hunted beaver and other animals along the streams before there were any other white settlers in this region. By taking a carriage one can enjoy a most delightful drive along the base of the Presidential Range, and from thence to Gorham, over what is known as the Cherry Mountain road. [See Gorham.]

JOBILDUNK RAVINE. See Warren.
JOCKEY CAP. See Conway.
JOHN'S PERIL. See Nahant.
JORDAN'S POND. See Bar Harbor.

K

KATAMA. See Martha's Vineyard.

KILLINGTON PEAK, one of the Green Mountain range, is seven miles east of Rutland, Vt., from which its summit is reached by a road nine miles long. The ascent is arduous, but the view from the summit, which is 3,924 feet high, is extremely fine. On the north side is Capital Rock, a perpendicular ledge 200 feet high. Near by are Mts. Ida and Pico and Castleton Ridge. Sutherland Falls are six miles north, where the Otter River plunges over a ledge of rock.

KING'S RAVINE. A tremendous hollow on the north of Mt. Adams, climbed for the first time in 1857 by Starr King, and named after him by the guides.

L

LACONIA, N. H. An attractive town on the Concord & Montreal Railroad, 102 miles from Boston. Here the tourist by this route catches his first glimpse of Mt. Washington, while passing aroung Round Bay, if the atmosphere is clear. The Belknap Range can also be seen upon the right while journeying north.

LAFAYETTE BROOK. See Franconia.

LAKE DUNMORE, a picturesque sheet of water at the foot of the loftiest range of the Green Mountains [see Green Mountains], and almost surrounded by bold hills, is reached by stage from Salisbury Station, Vt. (five miles), which is 27 miles north of Rutland. The lake is about four miles long, and a mile and a half wide at the widest part, and affords excellent boating, bathing and fishing. It is named in honor of the Earl of Dunmore, who visited it in 1770. On the east side of the lake, the massive and forest-clad peak of Moosalamos towers to the height of nearly 2000 feet and in a rugged ravine just beneath its crest is Llama Cascade, where a brook leaps down the mountain side in a series of picturesque cascades, which are visible from the lake. From the summit of Sunset Hill, on the west side, reached by a carriage road, a fine view of the Adirondacks, forty miles to the west, is obtained. There are many fine drives in the vicinity.

LAKE MEMPHREMAGOG, a beautiful sheet of water, 30 miles long and two to four miles wide, lying partly in Vermont and partly in Canada. It is reached by rail to Newport, Vt., 230 miles from Boston, via the Passumsic and connecting railways. Its shores are rockbound and indented with beautiful bays, between which jut out bold, wooded headlands, backed by mountain ranges. Numerous islands dot its surface. In ascending the lake by steamer from Newport. Indian Point, the Twin Sisters, and Province Island are passed. Then comes Tea-Table Island, and about half way down the lake is a landing from which a footpath leads to the summit of Owl's Head, 2743 feet high. The view is very extensive, including the entire length of the lake, the White Mountains, Lake Champlain, Willoughby Lake and Mountain, the St. Lawrence River, and Montreal. Skinner's Island and Cave, said to have been the haunt of Uriah Skinner, "the bold smuggler of Magog," during the War of 1812, are also near by.

LAKE OF THE CLOUDS. A small body of water about two miles southwest of the summit of Mt. Washington, near the Crawford Bridle Path. [See Crawford Bridle Path.] It is 200 feet long, 100 wide, and 12 feet deep. Its elevation is 5100 feet, and it is the source of the Ammonoosuc River.

LAKE WINNIPESAUKEE. The largest body of water in the White Mountain region. There are two derivations of its name given, both from the Indian tongue. One is "The Beautiful Water in a High Place," and the other, "The Smile of the Great Spirit." They are equally appropriate. The lake lies in the two counties of Carroll and Belknap, and is very irregular in form. Its area, exclusive of its 274 islands, is upwards of 71 square miles, and the distance around its shores is 182 miles. It is about 25 miles long, and from one to seven miles wide. There are 10 islands, having each an area of over 100 acres; and one comprises over 1000 acres. Long Island, which is in the northern part of the lake, about midway between Wolfboro and Centre Harbor, has lately been added to the list of summer resorts. At the west end the lake is divided into three large bays; at the north is a fourth, and at the east end there are three others. The waters descend 472 feet on their way to the Atlantic, forming a rapid river of the same name as the lake, and emptying into the Merrimac. There are no very large streams flowing into the lake, and it is supposed that a large part of the water supply comes from sub-aqueous springs. The largest streams which feed the lake are the Merrymeeting and Smith's rivers at the southern extremity; the other tributaries being short brooks or the outlets of adjacent ponds. The waters of the lake are so clear that the fish which abound in it can be distinctly seen playing among the stones at the depth of many feet. While Lake Winnipesaukee is distinctively a mountain lake, yet it lacks almost all those wild, rough features of mountain scenery which usually characterize inland waters in mountainous regions. The shore seen from a distance appears, as it is, comparatively smooth and level, but the mountains rise high on all sides. The islands which dot its surface are covered with verdure, and are neither rocky nor precipitous. The route by the way of Winnipesaukee is the finest approach to the White Hills. It has been happily compared to a vast antechamber, from which you look up from the valley of the Saco to the towering peaks of the mountains. Many of the most prominent of these are to be seen, and even the far distant top of Mt. Washington is visible on a clear day, scarcely distinguishable from the white cloud it pierces. Across the lake rise the higher eminences of the Ossipee range, and far in the north are the soft blue peaks of the lower range of the great series of hills, — Chocorua, Whiteface, and the rest. A more delightful experience than a trip across the water among the beautiful islands can scarcely be imagined. It is attended with little danger even in a squall, for although there are few buoys or lights to assist the pilot on his devious course, yet so familiar is he with the waters and the undulating shores, that even in a dense fog the journey is made with comparative safety. On leaving The Weirs, the boat winds its way among the numberless islands that crowd the course. A fine view is had of Red Hill, which is soon lost behind an intervening island. An opening strait gives a fine view of Rattlesnake Island, and Mt. Belknap or Gunstock is to be seen on the right, and on the left the ever-present Ossipee. Five miles from The Weirs is Bear Island, which is nearly four miles long. Upon one of the wild and romantic islands, passed by the boat, resided in 1851 an aged spinster, familiarly known as "Aunt Dolly," who for years lived here, almost entirely cut off from the world, cultivating a small farm, and tending a few sheep, and occasionally rowing her own skiff to the main land.

LAKE WINNISQUAM. See East Tilton.

LANCASTER, N. H. The shire town of Coös county, 211 miles from Boston, on the Concord & Montreal Railroad. Coös is an Indian name signifying crooked, and was originally applied to that part of the Connecticut River upon which the towns north of it are situated. Lancaster itself is not mountainous, but is surrounded by high hills. The whole range of the White Mountains, the Stratford peaks, the dark masses of the Pilot Range, and the beautiful Lunenburg Heights are in full view from some of its streets. Israel's River empties itself into the Connecticut at Lancaster. The drives in the neighborhood on either side of the Connecticut are unsurpassed, probably, in New England. A carriage road leads from here to Jefferson, nine miles distant.

LEDGES, THE. See North Conway.

LENOX, MASS. A favorite summer resort of people from Boston and New York, noted for the singular purity and exhilarating effects of its air and for the beauty of its mountain scenery. Among the famous characters who have made it their resort are Fanny Kemble Butler, Nathaniel Hawthorne, and Henry Ward Beecher, all of whom have told of its beauties. Excursions can be made to the summit of Bald Mountain, which commands a very fine view; the Ledge, Richmond Hill, and Perry's Peak. This isolated summit is six miles from the town, over 2000 feet high, and overlooks a vast range of country from the Catskills to the Green Mountains.

LISBON, N. H. A picturesque village on the banks of the Ammonoosuc River, a station of the Concord & Montreal Railroad, 178 miles from Boston. Sugar Hill is the name of an elevated section of the town lying near Franconia. [See Franconia.]

LITTLE BELLE ISLAND. See Newfound Lake.

LITTLE SQUAM. See Asquam Lakes.

LITTLETON. One of the largest and most prosperous towns in northern New Hampshire, on the Concord & Montreal Railroad, 188 miles from Boston. It is pleasantly situated in the valley of the

Ammonoosuc, on both sides of the stream, and extends up the neighboring hillsides. The adjacent hills afford noble views of the high mountains, and there are many drives and rambles in the vicinity. The rapids of the Connecticut River, known as the Fifteen-Mile Falls, border the town for a stretch of thirteen miles, and Waterford, Vt., is but five miles distant. Littleton is a favored summer resort, owing to its beautiful location and being of easy access to the principal mountain resorts, either by rail, stage-coach, or carriage. Bethlehem is but five miles distant, and the drive over the mountain road is one of the most delightful that can be imagined. It is also reached by a branch of the railroad which leaves the main line at Wing Road, six miles from Littleton, and continues up the Ammonoosuc valley to Fabyan's. At Bethlehem Junction, four miles from the main line, connection is made with the narrow-gauge road which runs to Bethlehem Street

LIVERMORE FALLS. A beautiful fall of water, formed by the Pemegewasset River in making its way through a rocky defile, between two and three miles north of Plymouth, N. H., and 128 miles from Boston, on the road to Franconia. It is a station of the Pemigewasset Valley Branch of the Concord & Montreal Railroad, The State maintains a fish hatchery there.

LONESOME LAKE. A small body of water, up among the mountains, southwest from the south entrance to the Franconia Notch. [See Franconia Notch.]

LONG POND. See Sebago Lake.

LONG LAKE. See Southwest Harbor; also Bridgton.

LOOK-OUT LEDGE. See Randolph, N. H.

LOUGEE POND. See Alton Bay.

LOVELL'S POND. See Conway.

LOWELL'S ISLAND. See Marblehead Neck.

M

MIANTONONI HILL, See Newport, R. I.

MINISQUAM. See Asquam Lakes.

MILL BROOK CASCADE. See Thornton.

MIRROR, THE. See Georgianna Falls.

MOAT MOUNTAIN. See Conway.

MOOSE CHASM. See Grafton Notch.

MOOSELOCMAGUNTIC LAKE. See Rangely Lakes.

MOOSEHEAD LAKE, the largest inland body of water in Maine, lies among the northern hills, on the verge of the great Maine forest. It is reached from Boston via railroad or steamer to Bangor, and thence via the Bangor & Piscataquis Railway to Greenville at the foot of the lake. Another route is via the Maine Central Railroad to Newport [see Newport, Vt.], thence via Dexter Branch to Dexter, 15 miles, whence stages run to Greenville. Stages also run to the lake from Skowhegan on the Maine Central Railroad, 100 miles from Portland. The lake is 35 miles long, and at one point is 12 miles wide, but near the centre there is a pass which is not more than a mile across. It is 1023 feet above the sea, into which it empties through the Kennebec River. Its waters are deep, and abundantly supplied with fish. The most favorable time for visiting it, to avoid the ravages of the black fly, is from May 15th to June 15th, and from Aug. 10th to Oct. 10th. A steamer plies between Greenville and the other end of the lake, and the scenery on the way is exceedingly fine. On the west side Mt. Kineo overhangs the water with a precipitous front over 600 feet high. Its summit is easily reached, and from it the lake is visible from end to end. To the northeast Katahdin stands out in massive grandeur against the horizon, 5385 feet high. It is a strangely isolated and graceful peak, the ascent of which is very arduous.

MOOSILAUKE BROOK. See North Woodstock.

MOSSY GLEN. See Randolph, N. H.

MOULTONBORO, N. H., originally the home of the Ossipee Indians.

MOUNTAIN COLISEUM. See Tuckerman's Ravine.

MT. AGASSIZ. See Twin Mountain Station.

MT. ANNANANCE. See Willoughby Lake.

MT. BELKNAP. See Alton Bay.

MT. BOND. See Twin Mountain Station.

MT. CANNON. One of the Franconia range of mountains, 3864 feet high, on the southern face of which are the ledges which form the face of the Old Man of the Mountain. [See Old Man of the Mountain.] The mountain derives its name from a large flat rock near its summit, which, when viewed from below, bears a strong resemblance to a cannon. A footpath leads to a point on the summit above the Profile.

MT. CLEVELAND. See Twin Mountain Station.

MT. CLINTON, one of the range forming the Crawford Notch. Crawford Plateau [see Crawford Plateau] is at its base, and the Crawford Bridle Path [see Crawford Bridle Path] to the summit of Mt. Washington, passes over it. Its summit is 4320 feet above the level of the sea, and it belongs to the great range which extends from the Notch northeasterly to Mt. Madison.

MT. FRANKLIN, one of the White Mountain Range, 4904 feet above the sea level, the summit of which can be reached by the Crawford Bridle Path. [See Crawford Bridle Path.] It has an irregular, flattened peak, from which a fine view of Chocorua [see Chocorua] is obtained. On its eastern side it is almost perpendicular.

MT. GARFIELD. See Twin Mountain Station.

MT. GUYOT. See Twin Mountain Station.

MT. KATAHDIN. See Moosehead Lake.

MT. KEARSARGE, sometimes called Pequawket, three miles from North Conway [see North Conway], is 3,251 feet high. A bridle path extends from the foot to the top. From the summit the whole White Mountain Range is to be seen, with a fine and unobstructed view of the peak of Mt. Washington. The sharp peak of Chocorua [see Chocorua] with Moat Mountain, 3200 feet high, and Middle Mountain, 2700 feet high, in the foreground, can also be seen with great distinctness. The course of Saco River can be traced almost from its source through the intervales.

MT. KINEO. See Moosehead Lake.

MT. KINSMAN. One of the Franconia Range on the west side of the Franconia Notch [see Franconia Notch], 4200 feet high. On its western slope are the Bridal Veil Falls, Copper Mine Brook, and Mt. Kinsman Flume. [See Bridal Veil Falls.]

MT. LAFAYETTE. One of the Franconia range, 5259 feet high. The view from its summit is very fine, including the southern valley of the Pemigewasset, Mt. Washington in the east. Katahdin in the northeast, and the hills of Stratford in the north. Glancing around the horizon, the rounded summits of the Green Mountains and the peaks of Monadnock and Kearsarge are to be seen. Upon the mountain is a strange looking stone of large size and unusual form named by those who discovered it The Altar. [See The Altar.] From the north-

42 MT. MAJOR

erly gate of Franconia Notch [see Franconia
Notch] is a bridle path to the summit of Mt. Lafayette. It winds along the base of Eagle Cliff, and
in some places the ascent is very steep. The time
occupied in the journey up and down is about
five hours.

MT. MAJOR. See Alton Bay.

MT. MANSELL. See Southwest Harbor.

MT. MANSFIELD, the loftiest of the Green Mountain
Range [see Green Mountains], 4348 feet high, can
be ascended by carriage road from Stowe, Vt. [See
Stowe.] Its summit has been found to resem-
ble the upturned face of a giant, showing the
Forehead, the Nose, and the Chin. The Nose has
a projection of 400 feet, and the Chin is thrust forward 800 feet. The distance from Nose to Chin is
a mile and a half. The Old Woman of the Mountain is a remarkable profile of the mountain. She
leans back in her easy chair, her work lying in her
lap, while she gazes across the valley. The carriage ride up the steep roadway to the base of the
Nose is full of interest, the changing views affording a constant succession of new and beautiful
prospects. The mountain sides, up to near the
summit, are very heavily wooded, but glimpses
can be caught of deep ravines. At one place the
road crosses a bridge that spans a yawning chasm
in the mountain side. From the summit a steep
and ragged path leads up the Nose, from the top
of which the view is little if at all inferior to that
from Mt. Washington. To the east are the White
Mountains, 60 miles distant, lying low along the
horizon. The isolated and symmetrical form of
Mt. Ascutney rises on the southeast. Southward
are Camel's Hump and Killington Peak, with innumerable smaller elevations of the Green Mountain Range. Westward lies a wide expanse of lowland, with many sparkling streams winding about
among the farms and forests and villages, the city
of Burlington in the distance, and beyond them
the beautiful expanse of Lake Champlain, with the
blue ridges of the Adirondacks bordering the far-
thest horizon. On the northwest is the Lamoille
Valley, watered by the Lamoille and Winooski
rivers; and far northward are Jay Peak and Owl's
Head, the beautiful St. Lawrence, a score of other
mountain peaks, and Lake Memphremagog. In
clear weather the mountains near Montreal, 70

miles distant, can be seen with the naked eye.
The Chin is 350 feet higher than the Nose, and
may be easily ascended by a path two miles long.

MT. MONADNOCK, (Jaffrey, N. H.) There is probably not another mountain of its size in New England which is so bare as Monadnock. The whole
upper portion is a ledge, and it stands out very
clear against the sky in all its naked and massive
grandeur. The view from its summit, 3186 feet
high, is extensive and pleasing, and the climb up
its sides, while calling for some exertion, is not
severe.

MT. MONROE, one of the Presidential Range [see
Presidential Range], 5384 feet high, is inferior to
Washington rather in height than in symmetrical
beauty. It has two majestic peaks, and one of its
sides forms the wall of Oake's Gulf. [See Oake's
Gulf.] The Crawford Bridle Path [see Crawford
Bridle Path] passes along the southeastern side of
the mountain, several hundred feet below the
summit.

MT. MOOSILAUKE. The highest peak in New
Hampshire west of Mt. Lafayette. It has an altitude of 4811 feet, and owing to the fact that it sur-
passes by from 1000 to 1500 feet the surrounding
heights, affords a view which embraces all the
chains of the White Mountain group, and extends
into Maine, Vermont, and Canada. It is situated
chiefly in Benton, a little distance across the line
from Warren. [See Warren.] The name of the
mountain has often been corrupted into "Moose-
Hillock," but its title is of Indian origin, and was
derived from two Indian words,—moosi, signifying "bald," and auke, "place;" the letter l having been inserted for euphony. There is a carriage road from Warren to the summit,—a broad
plateau from which its Indian name is derived.
The summit is covered with mosses, Alpine daisies,
and mountain cranberries. Moosilauke, being entirely isolated, is one of the grandest of view points
in the mountain regions. Professor Guyot pro-
nounces the view the most extensive in New England, not excepting that from Mts. Washington
and Lafayette, over which he says it possesses
many advantages.

MT. MORIAH, 4700 feet high, the summit reached
by a foot path from Gorham, N. H. The ascent is
not hard, and the view from the top is of surpass-

ing beauty. The mountains and valleys of northern New Hampshire, Vermont. and Maine form one great panorama of grand scenery, and on the west the great White Mountain Range is seen to fine advantage.

MT. PLEASANT. One of the range forming the Crawford Notch, the summit of which is reached by a bridle path from the Crawford Plateau. [See Crawford Plateau.] It is 4764 feet high, and has a peculiarly rounded top which presents a beautiful appearance from a distance. On the northern side are to be seen the effects of immense slides, which are supposed to have occurred, like most of those among the mountains, in the memorable storm of 1826. It lies between Mts. Clinton and Franklin. For another mountain of this name see Bridgton.

MT. PROSPECT, formerly known as North Hill, in the neighboring town of Holderness, is between four and five miles from Plymouth, N. H. It is 2072 feet in height, and from its summit a fine view of the higher mountains and neighboring lakes is obtained. There is a good carriage road to the top.

MT. RESOLUTION. See Crawford Notch.

MT. SURPRISE, one of the spurs of Mt. Moriah [see Mt. Moriah], is about 1200 feet high. The summit, which is reached by a bridle path easy of ascent, affords an unobstructed view of the grandest portions of the White Mountain ridge. Nothing prevents the eye from looking down 1200 feet to the bed of the Peabody River, and up along the forests to the peak of Mt. Madison, the crest of Mt. Adams and the summits of Mts. Jefferson and Washington. Eastman says, in his White Mountain Guide, that "There is no other eminence where one can get so near to these monarchs, and receive such an impression of their sublimity, the vigor of their outlines, their awful solitude, and the extent of the wilderness which they bear upon their slopes."

MT. WASHINGTON. The crown of New England lifts its bare, weather beaten summit 6293 feet above the level of the sea. In the month of June, 1642, Darby Field, one of the hardy pioneers of Pascataquack (Portsmouth), accompanied by two Indians, made what is believed to be the first ascent of Mt. Washington. The following August he led a larger party to the summit. From that

time to 1774 only two ascents are recorded. Old
Abel Crawford, styled the "Patriarch of the
Mountain," and for whom the Notch is named, at
the age of 75 made the first ascent ever made of
Mt. Washington on horseback. The first bridle-
path to the summit was cut in 1821 by his son,
Ethan Allen Crawford, called the "Giant of the
Hills." This is still known as the Crawford Bridle Path [see Crawford Bridle Path], and is used
by pedestrians in walking from Crawford Plateau
[see Crawford Plateau] to the summit, a distance
of nine miles, over the summits of Mts. Clinton
and Franklin and the sides of Pleasant and Monroe. The ascent can be made from Fabyan's over
the Mt. Washington Railroad [see Mt. Washington
Railroad], by carriage from the east side of the
mountain, and on foot from the same side, through
Tuckerman's Ravine. [See Tuckerman's Ravine.]
A hasty outline of the view from the summit,
which is unequalled in extent, grandeur, and beauty by anything east of the Rocky Mountains, will
assist the visitor in identifying many points in the
landscape. Looking north, the Presidential Range
first attracts attention, the order of the mountains
being Clay, Jefferson, Adams, and Madison. At
the foot of Madison is the Androscoggin River,
the course of which can be distinctly traced to its
source in Lake Umbagog. The villages of Berlin
Falls and Milan are just over Mt. Madison. Beyond the Androscoggin is an extensive group of
mountains in Maine, Mt. Blue, a sharp peak near
Farmington, being the most noticeable. Ebene
Mountain, 135 miles away, has been recognized in
this direction, and is, probably, the most distant
point to be seen from here. Looking east, the first
mountains are the Carter Range, surmounted by
Carter Dome, and joined on the south by Mt.
Wildcat. Beyond the Carter Range are seen Mts.
Baldface, Eastman, Slope, Sable, and Doublehead;
then comes a long stretch of lowland, dotted with
lakes, interrupted by the long ridge of Mt. Pleasant, near Bridgton. Still further east are Sebago
Lake and the city of Portland, while the ocean
itself, between the latter place and the Isles of
Shoals, is sometimes seen, but is generally difficult
to distinguish. Looking south, over Tuckerman's
Ravine and the Ellis valley, North Conway appears among the broad intervales of the Saco. At

the left is the beautiful dome of Kearsarge, near
which is Lovewell's Pond. Walker's Pond is just
south of North Conway, and on the right the valley is bordered by the north and south Moat
Mountains. Next on the right are Ossipee and
Silver Lakes, and then comes the sharp peak of
Chocorua. A little farther to the right is Lake
Winnipesaukee, with Mt. Belknap beyond it; and
next is the Sandwich Range, including Passaconaway, Whiteface, Tripyramid, and Black Mountain
or Sandwich Dome. Far beyond the latter are
Monadnock and the Southern Kearsarge; and Wa-
chusett, near Fitchburg, Mass., is just at the right
of Whiteface. Mt. Carrigain is the next conspic-
uous point, and beside it is the similarly shaped,
but more graceful, Osceola. Looking over the
southern part of the great range is seen Mt. Willey, overhanging Twin ranges. Whiteface, in the
Adirondacks, 130 miles distant, can be seen on a
perfectly clear day. Everything on the summit
is suggestive of the terrible wars waged by the elements in this unprotected region. The few buildings that have been erected are anchored in position by massive chains and cables. The first house
built on the mountain, a small stone cabin near
the top, erected by Ethan Allen Crawford, was de-
stroyed by the great storm in which the Willey
family perished, in 1826. In 1852, J. S. Hull and
L. M. Rosebrook built what was known as the
Summit House, a small story-and-a-half structure,
constructed of rough stones with a sharp wooden
roof, held in place by chains. A year later the
Tip-Top House was built by Samuel F. Spaulding
& Co., and run as an opposition hotel. All the
lumber for these houses was brought over the bridle-path on horseback. In 1873, the new Summit
House was erected, and the old houses are now
put to other uses. The old Tip-Top House is
occupied as the printing office of "Among the
Clouds," a daily newspaper published through the
summer. Besides these buildings are the United
States signal station, an observatory 27 feet high,
an office and two stables belonging to the coach
company, and the engine house of the railroad
company. A mile east of the summit is Tuckerman's Ravine [see Tuckerman's Ravine], and a
little further on are Hermit and Glen Lakes. Two
miles southwest of the summit, near the Crawford

Bridle Path, is the Lake of the Clouds. [See Lake of the Clouds.] The ownership of the summit of the mountain has been the subject of protracted lawsuits. Negotiations, recently concluded, have taken the controversy out of the courts, and the summit of the mountain has become the property of the Mt. Washington Railway Company.

MT. WASHINGTON CARRIAGE ROAD. A descent of Mt. Washington by way of the carriage road to the Glen and Pinkham's Notch [see The Glen and Pinkham's Notch], a distance of eight miles, gives a pleasing variety to a mountain tour. The road is admirably constructed, and the trip with careful drivers is attended with no more danger than on an ordinary highway. The road overcomes an elevation of 4750 feet, and the average grade is 592 feet to the mile, or about one foot in eight. The steepest rise, which is near the base, is one foot in six. Starting down the mountain, past the monument erected to Miss Bourne, the shelving rock under which Benjamin Chandler of Wilmington, Del., perished in a storm on August 7, 1856, is to be seen. A little further below, and at the right of the descending road, the ledge is visible where Dr. Benjamin Ball of Boston, in 1855, passed two nights in the snow and sleet of an October storm, and was rescued just in time to save his life. Continuing the descent, the great Gulf of Mexico yawns on the left of the carriage road, which runs along the very edge of this great abyss. On the opposite side are Jefferson, Adams, and Madison, looming up so grandly and distinctly that distance seems annihilated. Four miles from the summit is the Ledge, which commands a series of fine views of the northern peaks of the range, and also of the valleys to the east. The road now enters the forest, and for the remainder of the journey to the base, only occasional views of the mountains are to be obtained. Emerging from the woods, and passing the toll-house, the Glen and Notch are reached.

MT. WASHINGTON RAILWAY. Previous to 1876, the only direct means of reaching the base of Mt. Washington from Fabyan's was by a carriage road which runs directly through the wilderness. In that year an extension of the Concord & Montreal Railroad was built to connect with the road to the summit. It is six miles in length, and it has the steepest grade of any surface road in the mountains, the average rise being 183 feet to the mile, and the steepest over 300. For its entire length the road runs through a dense forest, and gradually approaches an impenetrable mountain wall, with no outlet save by the narrow thread of steel which winds around the flank and up through the clouds to the very summit of Mt. Washington. The whole chain of peaks may be seen, from Clinton on the south to Madison on the north, with Clay, Jefferson, and Adams towards the east. At Marshfield, the terminus of the surface road, is a small hotel and the shops and houses of the road, all located on the side of the mountain, at an elevation of 2670 feet above the sea. From Fabyan's the road up the side of the mountain appears to rise directly from the earth, and creep perpendicularly up the side of the mountain. From the base to the summit it is three miles long, and rises 3606 feet, an average of one foot in four. It was projected by Sylvester Marsh of Concord, N. H., who secured a charter for it in 1858. Many surveys were made, but no route seemed practicable except the one selected, which is a spur of the mountain formerly traversed by the old Fabyan bridle-path. Work on the road was begun in 1866, and the lower section as far as "Jacob's Ladder," was opened to the public on August 24, 1866. The track was finished to the summit the following July. The most formidable obstacles had to be overcome, and the track for almost the entire distance is supported by trestle work. The most wonderful piece of engineering on the road is what is known as "Jacob's Ladder," a trestle 30 feet high and 300 feet long, where the grade is 1980 feet to the mile, or 13 1-2 inches to the yard. The structure takes its name from a steep ledge near this point ascended by the bridle path. The essential peculiarity of the railway is the cog-rail laid in the centre of the track, and into which work the teeth of a large cog-wheel on each driving shaft of the locomotive, thus moving the train up the mountain. The boilers of the engines are inclined forward, so as to be nearly horizontal on steep grades. Each engine has two sets of cylinders and driving wheels, to obviate any danger from possible breaking of machinery, and in case of accident a system of breaks and checks that can effectually hold the train motionless on the steepest grade or allow the car disconnected from the engine to descend at a snail's pace. The track is strongly built, and is constantly being inspected. More than 500,000 passengers have been carried over the road, and no accident has ever occurred. During July, August, and September, five engines, each driving but one car, are run, so that visitors can start in the morning, having two hours at the summit, including time for dinner, and return in season to reach any point in the mountains the same night; or can go up in the afternoon, and descend the next morning, thus viewing those most wonderful of sights, a sunset and sunrise on the summit. The track also affords a direct and comparatively easy means of reaching the summit on foot. The only really difficult place to be encountered is "Jacob's Ladder," and that can be avoided by leaving the track at what is known as the "Half-Way House," where the Fabyan bridle path can, with little difficulty, be followed almost to the summit. By leaving the base an hour before the train starts, or even half an hour, no danger of being overtaken by it need be apprehended, as it moves little faster than a pedestrian would walk, the running time from the base to the summit being an hour and a half, including two stops for water. The ascent and descent have been frequently made by ladies over the trestle-work, the descent being the most difficult feat of the two. Miss Anna Dickinson is credited with having walked from Fabyan's to the summit and returned in one day, a distance of 18 miles. A few years ago two ladies accomplished the remarkable and difficult feat of walking from the base to the summit and return during a pleasant forenoon. The wonderful views to be obtained in the course of the ascent of the great ravines which scar the sides of Mt. Washington and the neighboring mountains will amply repay one for the journey, even though it be accomplished with some discomfort on foot, and even though the summit should be wrapped in clouds. On looking back down the track after the start is made, the windings of the turnpike can be easily traced, and an occasional glimpse of the river, as it plays in and out of the forest, is obtained. With each step upward the view widens, and Cherry Mountain, Jefferson, and a broad expanse of plain

4

and mountain is unfolded to the view. Gradually the dense forests, chiefly of spruce and birch, which cover the lower elevations, are left behind, and at an altitude of about 4000 feet nothing but a belt of dwarf spruces is to be found. Above that limit the only remains of vegetation are the Alpine flowers, moss of numerous varieties, and coarse grass. A view over the south wall is obtained, and on reaching the main ridge between Clay and Washington, there appears on the left of the track the huge Gulf of Mexico, an immense amphitheatre or ravine, down which one can look for a distance of almost 1000 feet. A short distance beyond, the monument of Miss Lizzie B. Bourne, whose sad death at this spot on Sept. 14, 1855, has been so often told, is passed, and the summit reached.

Mt. WILLARD. One of the White Mountain Range, about 2570 feet high, overlooking the northern gate of Crawford Notch. [See Crawford Notch] The ascent can be easily made from Crawford Plateau [see Crawford Plateau], over a carriage road two miles long. From the summit are to be obtained perfect views of the conformation of the Notch and the great mountains which form its boundaries. The most interesting time to visit the top is in the afternoon, when the lengthening shadows creeping up the side of the mountains on the east give a marvellous effect. An exceedingly fine echo can be produced by the notes of a cornet. Near the summit is the Devil's Den [see Devil's Den], and not far away is the Flume, a narrow and deep gorge through which a brook finds its way to the Saco River.

N

NAHANT, MASS., 13 miles from Boston, reached by stage from Lynn, on the Eastern Division of the Boston & Maine Railroad, and by steamer from Boston, is situated on a bold promontory of the Atlantic Ocean, connected with the main land by ridges of sand and stone thrown up by the ocean, above which its extreme high point rises more than 150 feet. It has a hard beach and splendid surf for a mile and a half, and a fine driveway. Curious rock formations, caused by the action of the waves, are Pulpit Rock, which lies just by the lower eastern shore of the horseshoe, between the

Natural Ridge (an arch of rock spanning a narrow crevice), and Sappho's Rock, rising 30 feet above the water, which, at a little distance, appears like a pulpit with Bible and prayer book; Swallows' Cave, a gloomy cavern at the lower end of the eastern curve of the horseshoe, overhung by a dome of irregular strata; John's Peril, which fancy might take for a yawning fissure in one of the cliffs; Castle Rock, a savage natural fortress; Spouting Horn, Caldron Cliff, and Roaring Cavern.

NANCY'S BROOK. See Crawford Notch.

NANTUCKET, MASS. A quaint old town on an island of the same name, about 30 miles from Martha's Vineyard. [See Martha's Vineyard.] It is reached by steamer from New Bedford on the Taunton Division of the Old Colony Railroad system. The island is of an irregular triangular form, about 16 miles long from east to west, and for the most part from three to four miles wide. It was at one time the chief whaling port of the world. It is picturesquely situated, and the old-fashioned houses, and the paved streets with the grass growing in them, present a very odd appearance. The principal attractions are the Cliff at the North Shore, whence a wide view is had; the beaches of the South Shore, where the waves roll in grandly after a storm; and Sciasconset (pronounced Sconset), a quaint little fishing hamlet on the southeast shore of the island, seven miles from Nantucket. It is reached by a primitive railroad, the track of which is covered with sand a part of the year. On Sanokty Head, one mile north of Sciasconset, there is a lighthouse, and from the eminence on which it stands, the Atlantic Ocean is visible on all sides of the island.

NARRAGANSETT PIER, one of the most popular of seaside resorts, is situated in the town of Kingston, R. I., on the Narragansett Pier Railroad, at the mouth of Narragansett Bay, and is equally attractive for bathing or riding. The surf is light and the water deepens very gradually, which, with the absence of strong currents, renders it unusually safe. Interesting places in the vicinity are the Heights, reached from the beach by street cars, and Silver Lake, a picturesque spot. Excursions may be made to Newport, Rocky Point, Providence, and Marked Rock, a popular excursion place a few miles higher up the bay.

NEWFOUND LAKE. A beautiful body of water, seven miles long and three miles wide, lying in the towns of Bristol, Bridgewater, and Hebron, N. H. It is reached by the Bristol Branch of the Concord Division of the Boston & Maine Raiload from Franklin. It is a favorite resort of camping parties. The distance from Bristol to the principal camping place is between four and five miles, over a good road, ascending most of the way. The principal fish are chub, perch, pickerel, black bass, lake trout, and land-locked salmon. The lake is surrounded by very high hills, and there are three mountains at the northerly end. There are quite a number of islands, mostly covered with wood, supplying camping retreats; Little Belle Island being a favorite place. There is a good supply of both sail and row boats. The shores have all the chracteristics of the mountain lake, being bold and precipitous in some places, and at others low and sandy. Off Sugar Loaf Mountain, 180 feet of line is required for fishing, so deep is the water.

NEWPORT MOUNTAIN. See Mt. Desert.

NEWPORT, R. I., called the "Queen of American watering places," is situated on a declivity of the southwest shore of the island from which the state is named, facing the harbor on Narragansett Bay. It is 69 miles from Boston, and is reached by the Providence Division of the Old Colony Railroad system; also via Newport & Wickford Steamboat Company and Continental Steamboat Company. The home of the summer residents, known as New Newport, stretches away to the south with a great number of cottages and villas of the most costly and ornate character. In and around the city are many interesting and beautiful localities. The most notable among the artificial curiosities is the Old Stone Mill, claimed by some to have been built by the Northmen five hundred years before Columbus discovered America; but the weight of evidence appears to be in favor of the theory that it was erected by Gov. Benedict Arnold, who died in 1678, and who spoke in his will of "my stone-built windmill." It is situated in Tuoro Park, given to the city by Abraham Tuoro, a Hebrew, who was born there. Near the Old Mill is a fine bronze statue of Commodore M. C. Perry, who was a native of Newport. The State House is a venerabel old building, fronting on Washington Square in the centre of the city It was erected in 1742. In the Senate Chamber is one of Stuart's celebrated portraits of Washington. The old Perry mansion, occupied by Commodore Perry after his victory on Lake Erie, fronts on this square; as does the City Hall. Other objects of historical or antiquarian interest are the old Jewish Synagogue, in Tuoro Street, and the Cemetery close by. The synagogue was built in 1762, and up to the Revolutionary War was regularly opened for worship. Since then it has remained unused, though it and the grounds are kept in good order by a bequest of $20,000 left by Mr. Touro for that purpose. Other places of worship of equal antiquity are Trinity Church, which dates from the last century, and has an organ still in use presented to it by Bishop Berkeley (then Dean) during his pastorate there, 1729-31; First Baptist Church, 1638; Central Baptist, 1733. The Vernon House, corner of Clark and Mary Streets, was the headquarters of Rochambeau in 1780. There are three fine beaches at Newport, and the facilities for surf bathing are unexcelled. The First Beach is the most popular, owing to its being so situated that there is no danger from undercurrents. It is about half a mile from the hotels, and about a mile east of it is Sachnest Beach, which is used only by the more adventurous, as the breakers are very heavy. At the west end of the beach is Purgatory, a dark chasm, 160 feet long, from eight to 14 wide at the top, from two to 24 wide at the bottom, and 50 feet deep. During storms the waves rush through it with tremendous fury. Above the Third Beach, a long secluded strip of sand, are the Hanging Rocks, within the shadow of which Bishop Berkeley is said to have written his "Minute Philosopher." The Spouting Cave, reached by Bellevue Avenue, the grand drive of Newport, is a deep cavern running back from the sea into the rocky cliffs, into which the waves rush madly during a storm, and dash through an opening in the roof, sometimes to the height of 50 feet. The Glen is a quiet and sequestered spot, seven miles out on the Stone Bridge Road. The Pirate's Cave, four and one half miles from the city, and Miantononi Hill, one mile and a half, are favorite resorts. Lily Pond, the largest sheet of spring water on the island, is easily reached from Spouting Cave. Fort Adams, on

Brenton's Point, three and one half miles from the city, is one of the largest and strongest fortresses in the United States, mounting 460 guns. Opposite it, on Conanicut Island, are the ruins of an old circular stone fort, called the Dumplings. From its crumbling walls a fine view of the harbor may be obtained. Lime Rock, famous as the home of Ida Lewis, lies in the harbor beyond Goat Island. Brenton's Cove is approached by a causeway leading to Fort Adams, and affords the best view of Newport that can be obtained.

NEWPORT, VT., 230 miles from Boston, is beautifully situated at the head of Lake Memphremagog on what is known as Pickerel Point, and is reached by the Passumsic Division of the Boston & Maine Railroad. Prospect Hill, just south of the village, commands a fine view of the lake and surrounding elevations, prominent among which are Owl's Head, Mt. Elephantes, Mt. Orford, Jay Peak and Mt. Willoughby. From the summit of Jay Peak, 4018 feet high, 12 miles west of the village, a fine view is obtained of the entire range of the Green Mountains, including Mt. Mansfield and Camel's Hump, Killington Peak, Ascutney Mountain, White and Franconia Mountains, Mt. Kearsarge, Lake Champlain, and the Adirondacks. Other places of interest are Clyde River Falls (two miles), Bear Mountain (seven miles), and Bolton Springs, Canada, (15 miles).

NORMAN'S WOE. See Gloucester.

NORTH CONWAY, N. H. A beautiful village in the Saco Valley, surrounded by mountains, 138 miles from Boston, 60 from Portland, and 31 from Fabyan's on the Eastern Division of the Boston & Maine and Maine Central railroads. In the east are to be seen the peaks of Kearsarge, on the west the Moat, shutting in the valley, and in the north is the Presidential Range. The summit of Kearsarge, 3251 feet high, is reached by a bridle-path, and a foot path extends to the top of Moat Mountain. The views from these mountains are very fine. Among the principal places to be visited in the neighborhood are the Artist's Falls, more remarkable for their beauty than greatness; Echo Lake, the Cathedral, and the Ledges, all situated on the opposite side of the river, and about three miles distant. Echo Lake is a small sheet of water with a remarkable echo, lying at the base of Moat

Mountain. There is a series of cliffs extending for four or five miles along the mountain side, varying in height from 100 to 800 feet, in one of which the Cathedral is situated. This is a cavity in the solid granite rising some 80 feet, roofed with solid rock, and having a floor about 20 feet in width. Diana's Bath, a little farther to the north, is a series of basins in the rock, supplied with water by a beautiful cascade above. Pictured upon the perpendicular sides of the cliffs, as seen from the village, is the figure of a horse.

NORTH WOODSTOCK. The northern terminus of the Pemigewasset Valley Branch of the Concord & Montreal Railroad. The village is 21 miles from Plymouth, and 150 from Boston. It is within about ten miles of the Profile House, with which it is connected by stage. The valley here is wide, and there are numerous side valleys, a multitude of trout brooks, and mountain views on all sides. The east branch of the Pemigewasset River enters here, coming from a pass through the mountains which leads direct to Fabyan's. There is a foot path through this valley, leading to numerous good trout streams. From the west flows the Moosilauke Brook, on which is the Agassiz Basin, where is some remarkable rock scenery. Through this valley a path eight miles long leads to Moosilauke. [See Mt. Moosilauke.] There is a point on this path where nearly all the Presidential Range and many other grand mountains can be seen. Between one and two miles from North Woodstock is Georgianna Falls. [See Georgianna Falls.]

O

OAK BLUFFS. See Martha's Vineyard.

OAKE'S GULF. A great ravine formed on one side of Mt. Monroe [see Mt. Monroe], and on the other by mountains of the White Mountain Range. [See White Mountain Range.] Its walls are a series of perpendicular craggy precipices, and its bottom is covered with huge rocks scattered in wild confusion. The Crawford Bridle Path [see Crawford Bridle Path] passes along its southwestern side on Mt. Monroe.

OGUNQUIT BEACH. See York Beach.

OLD MAID OF THE MOUNTAIN. See Crawford Notch.

OLD MAN OF THE MOUNTAIN, THE. A remarkable resemblance to the human face to be seen in the Franconia Notch [see Franconia Notch] about half a mile from its northerly gate. It is formed by a series of three ledges on the southern face of Mt. Cannon [see Mt. Cannon], nearly 1500 feet above the lake at its base, one of which forms the forehead, another the nose and upper lip, and the third the chin. The brow is massive and projecting, the nose straight, finely cut and sharply outlined, the lips thin and senile and slightly parted, and the chin is well thrown forward, with exact proportionate length. The length of the Profile is from 60 to 80 feet. As a whole it is symmetrical, and, as seen from one point in the valley, perfectly distinct and clear. When viewed from the front, however, all resemblance to a human face is lost, and it is only at the place where a guideboard has been erected by the side of the road that the Profile is to be distinctly seen. It was discovered in 1805 by two men at work on the Notch road, and since then has been an object of absorbing interest. The precipice, of which it forms the extremity, is not unlike the Palisades of the Hudson in appearance. It extends for nearly two miles along the escarpment of the mountain, and is a prominent part of the scenery of the section.

OLD ORCHARD BEACH, four miles from Saco, Me., reached by stage, or by trains on the Western Division of the Boston & Maine Railroad, is the finest beach in New England, and, after Swampscott and Rye, the most frequented and fashionable. It is nearly 10 miles long, is hard and smooth as a floor, shelves gently to the water, and affords unsurpassed surf bathing. On Foxwell's Brook is a waterfall of 60 feet, surrounded by wild and romantic scenery. The fishing and shooting in the vicinity are excellent.

OLD WOMAN OF THE MOUNTAIN. See Mt. Mansfield.

OQUOSSOC LAKE. See Rangeley Lake.

OSGOOD'S FALLS. See The Glen.

OWL'S HEAD. See Lake Memphremagog.

P

PEMIGEWASSET MOUNTAIN. A high spur of Mt. Kinsman. [See Mt. Kinsman.] From its summit a fine view is obtained of the superior peaks on the opposide side of the Franconia Notch and of the valley southward.

PEMIGEWASSET VALLEY. A beautiful stretch of country extending on either side of the Pemigewasset River from Plymouth, N. H., to the Franconia Notch. In it are the towns of Campton, Thornton, and Woodstock. It has always been a favorite resort for artists and tourists. The approach to the mountains of the Franconia range is full of interest, furnishing, as it does, a constant succession of beautiful landscapes. The valley in many places is broad, and the intervales are rich and fertile. The river flows placidly at times through green and luxuriant meadows, and then in rapid and headlong torrent over a pebbly bed. The dark hills rise on either side, and in the distance are the bold outlines of Mts. Lafayette, Lincoln, Liberty, Flume, Pemigewasset, Cannon, and Kinsman. A turn in the road reveals a most charming picture, and the broad outline of the mountain forms about the Franconia Notch become more and more distinct. At Campton, the view opens to the right up the Mad River Valley, completely shut in by dark hills. The upper part of the valley is mainly a wilderness, with but few houses until the Flume and Profile Houses are reached. Beyond North Woodstock [see North Woodstock] the valley narrows, and the dark mountains close in upon either side, with here and there a little intervale of field and meadow, with miles of forests beyond. Little Coolidge, Big Coolidge, and the Potash mountains rise upon the right, Pemigewasset Mountain on the left, while in front, apparently obstructing further progress in that direction, are the huge forms of Flume, Liberty, Lincoln and Lafayette. Here begins the Franconia Notch. [See Franconia Notch.]

PERRY'S PEAK. See Lenox.

PHILLIPS BEACH. See Swampscott.

PIGEON COVE, 33 miles from Boston, on the Eastern Division of the Boston & Maine Railroad, is situated at the extreme point of Cape Ann. By reason of the great beauty and sublimity of its scenery, the healthfulness of its climate, its medicinal springs, its splendid surf and still-water bathing, it is a much frequented summer resort.

PINKHAM NOTCH. A pass about 10 miles in length, running between Mts. Washington and Carter from

the Glen [see The Glen] to Jackson Village. [See
Jackson.] Starting from the Glen, the carriage
road follows closely the Peabody River, passing
Emerald Pool on the right, a quiet basin in the
river which is in strangs contrast to the tumultu-
ous water that enters it. Then comes Thompson
Falls [see Thompson Falls] a little beyond, on the
left. About two or three miles from the Glen, the
road crosses the river twice in quick succession,
and reaches the highest point of the Notch. The
Peabody and Ellis Rivers here issue from the forest on the right in nearly parallel courses, and
very near to each other. The Peabody turns abruptly down the ravine to the north to unite with
the Androscoggin River, while the Ellis River
takes the opposite course down the Notch towards
the Saco River. Between these two streams is the
entrance to the Crystal Cascade [see Crystal Cas-
cade], and a little further on, on the left, is the
path to the Glen Ellis Falls. [See Glen Ellis Falls.]
Passing down the Notch for two or three miles, an
opening gives a fine view of Mt. Washington. It
was near here that Captain Joseph Pinkham settled in April, 1790, coming up over the snow with
his family, bringing all their household goods on
a handsled. Daniel Pinkham, one of his sons, constructed the road through the Notch which bears
his name. This is really the foot of the Notch, and
Jackson is some miles beyond.

PLEASANT MOUNTAIN. A line of wooded heights in
Bridgton [see Bridgton] and Denmark, Me., between the Saco Valley and the Morse Ponds. It
consists of several rounded crests separated by
shallow ravines, and from distant points presents
the appearance of a long wall. On account of its
isolated position, its summit, 2018 feet above the
sea, is one of the best points from which to obtain
a distant view of the outlines of the great White
Mountain Range. Some 50 lakes and ponds may
be distinctly seen with the naked eye. Looking
eastward are to be seen Moose Pond, Wood's Pond,
Highland Lake, Long Lake, the Bay of Naples,
the Harrison and Otisfield hills, and the villages
of Bridgton, North Bridgton and South Bridgton.
Southward are Saddleback Mountain, Mt. Cutler,
Sebago Lake, and Portland. In the west are Brown-
field and the Ossipee Mountains. In the northwest, the Saco River, Lovewell's Pond, Pleasant

Pond, Round Pond, Kezar Pond and River, Jockey
Cap, Oak Hill, Fryeburg Village, and Mt. Choco-
rua; and further north, Kearsarge and the White
Mountain Range. In the northeast, Waterford
Village, Bear and Hawk Mountains, Norway, and
Paris Hill.

PLYMOUTH, N. H. One of the most beautiful towns
in New England, on the Concord & Montreal Railroad, 126 miles from Boston. The Pemigewasset
Valley Branch diverges from the main line at this
point. Plymouth is delightfully situated on a ter-
race above the west bank of the Pemigewasset
River at a point where the valley widens into broad
intervales, to which a veritable forest of elms gives
a cool and inviting appearance. One of the principal objects of interest in the village is the old
court-house in which Daniel Webster made his
first plea before a jury. It originally stood in the
south part of the town, but in 1875 it was removed
to its present position in the rear of the new court-
house, and converted into a library building. It is
a small, one-story structure, and has a quaint and
antiquated appearance. There are many charming drives in the vicinity of Plymouth, and probably none of the mountain towns presents greater
attractions in this way. Near by, at Livermore
Falls [see Livermore Falls], is the fish-hatching es-
tablishment belonging jointly to the States of New
Hampshire and Massachusetts; and five miles distant is Mt. Prospect. [See Mt. Prospect.] There
are three routes from Plymouth to the White
Mountains, viz: the Concord & Montreal Railroad,
which runs along the east side of the great range;
the stage coach, in which a delightful ride of 29
miles carries the traveller completely through the
Franconia Notch; or the Pemigewasset Valley
Railroad, which affords an opportunity of making
the trip along the banks of the river through a
perfect wilderness to North Woodstock.

POND OF SAFETY. A small body of water 2000 feet
above the level of the sea, high up among the
mountain ridges at Jefferson, N. H. It is the chief
source of the Upper Ammonoosuc River.

POOL, THE. A famous and wonderful freak of
nature near the southerly gate of the Franconia
Notch [see Franconia Notch], and about half a mile
east from the main road. A path through the
woods leads directly to it. The Pool is a deep ex-

cavation in the granite as though hewn by human hands, and holds the waters of Pemigewasset River, here a small stream, which enter by a cascade from the upper extremity, and escape through an opening in the mass of rocks at the lower side. The width of this formation is about 140 feet, and its depth about 40 feet. The distance from the brink of the wall above to the surface below is nearly 150 feet. The presiding genius for years was a man who apparently lived in a boat of somewhat novel construction that floated on the waters of the Pool, and who for a small sum would give any one who desired a ride in his ark-like dwelling. He had a wonderful degree of confidence in the healthfulness of his habitation, and boasted that during the many summers he had lived in this secluded spot he had never had a cold. He constructed a path from the Pool to the Flume which considerably shortens the distance between the two places.

PORCUPINE ISLANDS. See Bar Harbor.

POTTER PLACE. See Mt. Kearsarge.

PRESIDENTIAL RANGE. The six mountains in the White Mountain Range [see White Mountain Range] named for Presidents of the United States. They are Washington, Jefferson, Adams, Jackson, Madison, and Monroe.

PROFILE LAKE. A beautiful sheet of water at the foot of Cannon Mountain in the Franconia Notch. [See Franconia Notch.] It was formerly known as Ferrin's Pond.

PROFILE MOUNTAIN. See Cannon Mountain.

PROFILE, THE. See Old Man of the Mountain.

PROSPECT HILL. See Alton Bay.

PULPIT ROCK. See Nahant.

R

RAFE'S CHASM. See Gloucester.

RAGGED MOUNTAIN. See Andover.

RANDOLPH. A village six miles from Gorham, and nine miles from Jefferson, located among the northern peaks of the White Mountains, 1200 feet above sea level, on the main road from Jefferson to Gorham [see Jefferson and Gorham], which passes over the crest of a hill 600 feet higher than the village, from which eminence may be obtained excellent views of Mts. Madison and Adams, and the remarkable gorge of King's Ravine. [See King's Ravine.] There are walks and drives from Randolph to Mossy Glen, Ice Gorge, Salamacis, Cold Brook, Triple Falls, Pond of Safety, Look-out Ledge, King's Ravine, Jefferson Highlands, Crystal Cascade, Glen Ellis Falls, and the summit of Mt. Washington.

RANDOLPH HILL. See Gorham.

RANGELEY LAKES, situated on the north west coast of Maine, within the borders of its great forest region, comprise a chain of picturesque bodies of water, connected by narrows and streams, extending from the Oquossoc or Rangeley Lake, 1511 feet above the sea, to the Umbagog, 1256 feet above the sea, forming one continuous water way for a distance of nearly 50 miles, and embracing 80 square miles of water surface. Each lake has its individual name, but the chain is known collectively by the title given above. The lakes can be reached by the Maine Central Railroad to Farmington, 83 miles from Portland; thence the Sandy River Railroad to Phillips; and from there by Phillips & Rangeley Railroad to Rangeley, a distance of 29 miles. Indian Rock, near by, is a favorite old Indian camping ground, and is a headquarters for sportsmen, being the most central point in that region, and within half a mile of the great Mooseloemaguntic and Catsuptic Lakes. All the waters of this region abound in fish, and the forests in game.

RAYMOND'S CATARACT. See The Glen.

RED HILL. A mountain situated in the town of Moultonboro, about four miles distant from Centre Harbor [see Centre Harbor], from which there is a good wagon road to the base. It is 2038 feet high, and the summit is reached by a bridle path. It is by no means difficult to climb. The top, being destitute of trees and bushes, affords an uninterrupted prospect of the lake and distant mountains. In clear days the peaks of the White Mountains are discernible in the distant north, the Ossipee Mountains are visible in the east, a little to the north is Chocorua, and still farther away are to be seen the mountains of Maine. Kearsarge and Monadnock are plainly visible at the southwest, and Belknap at the southeast. Squam Lake, dotted with beautiful green islands, fringed with beaches of white sand, adds its manifold charms to the

view in the west. To see Lake Winnipesaukee at
its best, the ascent of Red Hill should be made in
the very early morning, or in the afternoon. The
rising and the setting sun gives a peculiar charm
to this most charming of lakes, and, with the ad-
vancing and receding light, the shadows of the
hills rise and fall on its surface, forming myriads
of weird and fantastical figures. The mountains
on the opposite shore in the afternoon change
from a glow of crimson to a brown purple, intro-
ducing, with gorgeous effect, all the intermediate
tints. Coaches leave Centre Harbor for Conway
and North Conway regularly every day, soon after
the arrival of the morning boats from the Weirs
and Alton Bay.

RIDGE ROAD. See Conway.

ROARING CAVERN. See Nahant.

ROCKPORT. 35 miles from Boston, on Cape Ann, is
a popular seaside resort. It is reached by the Cape
Ann Branch of the Eastern Division of the Boston
& Maine Railroad. Fine bathing and grand sea
views are its chief attractions. At the entrance to
the harbor is Thatcher's Island, on which are two
famous lighthouses.

ROCKY BRANCH, a small mountain stream in Crawford Notch [see Crawford Notch] which empties
into the Saco River [see Saco River] near Upper
Bartlett. [See Upper Bartlett.] It is fed by the
mountain rivulets, and its banks overflow frequently in the spring. At the time of the Willey
Slide, described under the title Crawford Notch,
it rose so rapidly as to surround a log cabin on its
banks before the inmates could make their escape.
It was floated down the stream, but grounded on
the summit of a little hill, where the frightened
family landed in safety.

ROCKY POINT, R. I., famous for its clam bakes, is
situated on Narragansett Bay. On the summit of
a hill is an observatory 125 feet high, from which
an extended view of the bay is to be obtained.

RUMFORD FALLS. See Bethel.

RYE BEACH, the most fashionable of the New
Hampshire beaches, is reached by a delightful
drive of seven miles from Portsmouth, 56 miles
from Boston, on the Eastern Division of the Boston
& Maine Railroad, or by stage from North Hampton station. The surf is particularly fine and
without any undertow. From Straw's Point, near

by, a grand view is obtained, including the Isles
of Shoals [see Isles of Shoals], and a great extent
of coast line.

S

SACO LAKE. See Crawford Plateau.

SALISBURY BEACH, N. H., one of the best on the
coast, extends about six miles from the mouth of
the Merrimac River, at Newburyport, 36 miles
from Boston, on the Eastern Division of the Boston
& Maine Railroad, to the Hampton River, and is
so firm and hard that a horse's hoofs make hardly
any impression. During the summer months it is
a lively and popular resort. The shore descends
very gradually, and the bathing is excellent.

SANBORNTON BAY. See East Tilton.

SANKOTY HEAD. See Nantucket.

SAPPHO'S ROCK. See Nahant.

SAWYER'S ROCK. See Crawford Notch.

SCIASCONSET. See Nantucket.

SCREW-AUGER FALLS. See Bethel.

SEA VIEW BOULEVARD. See Martha's Vineyard.

SEBAGO LAKE. A station of the Maine Central
Railroad, 17 miles from Portland, Me., on Sebago
Lake. The Adventists hold campmeetings here,
and it presents many attractions in the way of excursions on the lake, boating, and fishing.

SEBAGO LAKE, a beautiful body of water in Cum-
berland County, Me., on the line of the Maine
Central Railroad, 17 miles from Portland, and 74
miles from Fabyan's. Its name is of Indian origin,
meaning a "stretch of water." It is 14 miles long
by 11 wide, and receives the waters of 23 ponds.
There are but few islands in the lake, the greater
part of which is an unbroken expanse of water,
flanked by low shores, from which rise gracefully
curving ridges. Leaving Sebago Lake Station by
steamer, the tourist passes Indian Island, with an
area of 75 acres, and Frye's Island, with its thousand acres of dense forest. Sailing up the eastern
shore, the Notch, a narrow neck of water between
the island and Raymond Cape, is entered. Below,
on the right, are the Images, a curious mass of
rocks rising perpendicularly from the water nearly
70 feet, and then sloping in jagged, fanciful shapes,
to a further height of some 30 feet. Here, too, is
the Cave, which possesses a peculiar interest from
the fact that it was a favorite boyhood haunt of

Nathaniel Hawthorne. It is a square aperture, four feet by six, in the solid rock, into which the great novelist was wont to sail in his fishing boat to a distance of 25 feet, and then clamber through a short passage to the outer world. To the north-west is to be seen the early home of Hawthorne. The scenery on the west is wilder and more rugged. Saddleback Mountain, in Baldwin, is plainly visible, from which the eye roams northeast to Peaked Mountain, beyond which the view extends northward to Mt. Kearsarge and the White Mountain Range. At its northwestern end the lake connects, by the Songo River, with Long Pond, a river-like body of water nearly 14 miles long, and only two miles wide. The distance between the two lakes is but two and a half miles, but the Songo, the crookedest of all rivers, makes 27 turns and covers a distance of six miles. The passage up the river is the most interesting part of the trip. Five miles from its mouth is the Lock, by which the steamers and other craft plying upon these waters are raised from the level of the lower to the upper lake. One mile above is the Bay of Naples, two miles long, and Chute's River, a short stream, connects it with Long Lake. From there to Bridgton [see Bridgton], the distance is nine miles, and the places of interest passed are Long Point, Bear Point, Lovejoy's Island, Mast Cove Landing, Pleasant Point, and Mt. Henry. The steamer continues on a few miles northward to the pretty village of North Bridgton, and across the head of the lake to Harrison Village.

SHEEP MOUNTAIN. See Alton Bay.

SILVER CASCADE. A perpendicular fall of a small stream of water for nearly 400 feet in the Crawford Notch [see Crawford Notch] just below the northern entrance. It glides over the surface of the ledge above in an unbroken sheet of water after a heavy rain, concealing with spray a huge rock just below the summit, which during a drought divides the current. At first the water is diffused over a broad surface, but, before reaching the base, it is compressed into a very narrow channel. From the bridge which crosses the stream at the base of the mountain, a fine view is to be had of this remarkable fall.

SMUGGLER'S NOTCH. A wild and picturesque pass between Mts. Mansfield [see Mt. Mansfield] and

Stirling. It is eight miles from Stowe, Vt. [see Stowe], from which it is reached by a very good road. The sides of the Notch rise to an altitude of 1000 feet, the upper verge of the cliffs towering above the fringe of trees on their sides. The floor is covered with immense boulders and fallen masses of rocks, covered with mosses and ferns. Even great trees have found nourishment in the crevices between the rocks in this dimly lighted vault, their roots encircling the huge boulders in their search for soil and moisture.

SMUTTY NOSE ISLAND. See Isles of Shoals.

SNOW ARCH. A cave formed by the action of the water on the snow in Tuckerman's Ravine. [See Tuckerman's Ravine.] The snow is blown over from the summit of Mt. Washington by the northwest winds in winter, forming a drift of a hundred feet deep under the arched walls of the ravine. The warm streams from the mountain during the spring and summer tunnel this bank till they form an arch under which a person can easily walk. A measurement of this Arch in August, 1855, showed it to be about 300 feet long, 70 feet broad, and 15 feet deep. The roof was five feet thick, and so solid that it was with difficulty cut with a hatchet. The bank usually remains till the last of August.

SNOW CAVE. See Dixville Notch.

SOMES' SOUND is an inlet extending nine miles into the southern extremity of Mt. Desert Island. It is deep enough for an ocean steamer, but so narrow that from the deck of the Golden Rod, that plies between Somesville at its head and Bar Harbor, the voices of the men at work on the great stone ledges on either side can be distinctly heard, the stroke of their hammers making a musical accompaniment to the motion of the steamer. There are striking views on either hand. Eagle Cliff rises perpendicularly to a height of nearly 1000 feet on Eagle Mountain, and Fernald's Point, on the west side of the Sound, is the site of the ancient Jesuit settlement of St. Sauveur, near which is Father Biard's Spring. The Sound affords excellent fishing and boating.

SOMESVILLE. A favorite resort at the head of Somes' Sound, Mt. Desert.

SONGO RIVER. See Sebago Lake.

SPOUTING HORN. See Nahant.

5

SORRENTO. One of the most charming spots along
the Maine coast. It is just six miles across from
Bar Harbor [see Bar Harbor], and was originally a
part of Sullivan. It was named by the late Rev.
H. Bernard Carpenter, who was quick to note the
resemblance in its situation to that beautiful city
on the Bay of Naples. There are many fine summer residences and a spacious hotel along the terraces that edge Frenchman's Bay, and by driving
a few miles one is immediately in a thickly wooded
country which tempts the sportsman by its famous
trout streams. Tunk Pond, a few miles inland, is
one of the many attractive camping places of the
locality.

SOUTHWEST HARBOR is on the southwest side of
Mt. Desert, an island in Frenchman's Bay, just off
the coast of Maine, about 110 miles east of Portland,
and 40 miles southeast of Bangor, reached by the
main line of the Maine Central Railroad to Bangor, or by the boats of the Boston & Bangor Steam-
ship Company, and thence by the Mt. Desert
Branch to Bar Harbor Ferry Station. The island
is 14 miles long, and eight miles wide at its widest
part. At its northern end it approaches so near
the main land that they are connected by a bridge.
Nearly midway it is pierced by an inlet known as
Somes' Sound [see Somes' Sound], which is seven
miles long. In his "Summer Cruise," Mr. Carter
says of Mt. Desert: "The island is a mass of
mountains crowded together, and seemingly rising
from the water. As you draw near they resolve
themselves into 13 distinct peaks, the highest of
which is about 2000 feet above the ocean. Certain-
ly only in the tropics can the scene be excelled."
The mountains are mainly upon the southern half
of the island, and lie in seven ridges running nearly
north and south. The highest peak is Green
Mountain, and the next, separated from it by a
deep, narrow gorge, is called Newport Mountain.
The western sides of the range slope gradually
upward to the summits, but on the east they con-
front the ocean with a series of stupendous cliffs.
High up among the mountains are many beautiful
lakes, the largest of which is several miles in
length. These lakes, and the streams that flow
into them, abound in trout. There are several
summer resorts on the island, the best known of
which are Southwest, Northeast and Bar Harbor.

[See Bar Harbor.] Southwest Harbor, while less picturesque in its surroundings than the eastern and northern shores, has several points of interest. Chief of these is the sea wall, three miles southwest, a mass of shattered rock skirting the shore for a mile, against which the sea beats with tremendous force. Beach, Dog, Flying, Mansell, and Sargent's mountains may all be ascended from here affording fine views. Long Lake is about two miles northwest; Denning's Lake about three miles north, and Seal Cove, five miles west. The scenery is fine, and the lakes abound in fish.

SQUAM LAKE. See Asquam Lake.

STAR ISLAND. See Isles of Shoals.

STOWE, VT. A small town reached by stage from Waterbury, 12 miles north, delightfully situated on a plain, surrounded by noble mountain scenery. From Sunset Hill a fine view is obtained of the village and its natural attractions. Among the favorite excursions are those to the summit of Mt. Mansfield [see Mt. Mansfield], Moss Glen Falls (three miles), and Gold Brook (three miles). Next to the ascent of Mt. Mansfield, the great attraction is Smuggler's Notch. [See Smuggler's Notch.]

STRAW'S POINT. See Rye Beach.

SUGAR HILL, N. H. See Lisbon.

SULLIVAN HARBOR has undoubtedly the best view of the magnificence of Mt. Desert scenery of any neighboring resort. It is just a pleasant sail from Bar Harbor, and during the summer steamers run frequently. One of the features of this place is its great elm trees that shade the natural terraces that form its roads, all running parallel with Frenchman's Bay. Sullivan is on the main land, and its drives in all directions are a delight in all seasons.

SUNAPEE LAKE. A beautiful sheet of water in a basin on the height of land which divides the waters of the Connecticut and Merrimac rivers, and borders the eastern part of Sullivan and the western part of Merrimac counties, N. H. It is surrounded by the towns of Newbury, New London, and Sunapee. The name Sunapee is derived from the Algonquin words Suna and apee, meaning goose-water, and was given to it because it was a favorite resort for wild geese, which gave it an additional attraction to the Indians. It is about 10 miles in length, and from one half to one and a half miles in width. High hills and mountains surround it on all sides,—Sunapee, Croydon and Grantham, and grand Kearsarge being among its mountain sentinels, while Ascutney, in Vermont, is in sight over the lower western elevations. Numerous beautiful wooded islands add charm to the surface, while the irregular shape of the lake gives many projections of land and indentations of water, supplying the most favorable locations for cottages and camping places. The west shore is generally bold, while on the east shore there are several beaches of fine white sand.

SWALLOWS' CAVE. See Nahant.

SWAMPSCOTT, a favorite seaside resort of the wealthy people of Boston, is 12 miles from that city on the Eastern Division of the Boston & Maine Railroad. The shore is lined with elegant villas surrounded by beautiful grounds. There are three beaches, and picturesque headlands reach out into the sea. The bathing is excellent, with no undertow, and the water is said to be warmer than at Nahant or Rye Beach. Between Swampscott and Marblehead are Phillips Beach, Beach Bluff, and Clifton, all popular summer resorts.

SYLVAN GLADE CATARACT. A wild and beautiful waterfall on Mt. Willey, about two miles above its base. It is formed by a small brook that empties into the Saco River near the Willey House. [See Crawford Notch.] The water flows between the granite walls of a very steep ravine, and leaps first over four rocky stairways, each of them about six feet high, and then glides, at an angle of about 45 degrees, 150 feet with many graceful curves down a solid bed of granite into a pool below. The cascade is about 75 feet wide at the base and 50 at the summit.

T

THOMPSON FALLS. A picturesque series of cascades formed by a brook, a tributary of Peabody River, about two miles from the Glen [see The Glen] on the road to Jackson through Pinkham's Notch. They are about a quarter of a mile from the road by a path leading into the woods. By following the brook up the mountain side for half a mile, the last of the series is reached, and a fine view obtained of Mt. Washington and Tuckerman's Ravine.

THORN MOUNTAIN. See Jackson.

THORNTON, N. H. A village on the Pemigewasset Valley Road, nine miles from Plymouth and 135 from Boston. It presents many rare attractions to the summer sojourner. Mill Brook Cascades are in this town, and are visited either from here or from Campton Village.

THOUSAND STREAMS. See Tuckerman's Ravine.

TIN MOUNTAIN. See Jackson.

TUCKERMAN'S RAVINE. A tremendous gulf in the southerly side of Mt. Washington, named in honor of Edward Tuckerman, a botanist, who often visited the Ravine to obtain information of the ferns and lichens of the region. The ravine may be reached by continuing up the mountain by the path that leads to the Crystal Cascade [see Crystal Cascade], or by following the Mt. Washington carriage road two miles from the Glen [see The Glen], striking off into the forest by a path marked by a guideboard, or descend into the Ravine from the summit of Mt. Washington, a distance of a mile. All of these routes are attended with difficulties, requiring good power of endurance, the journey up the mountain being by precipitous paths covering a distance of about five miles. In approaching the Ravine from below, one passes Hermit Lake, a little sheet of water so completely isolated that the name given it is at once recognized as a fitting one, and is confronted by the great wall of the Ravine that looms over it. The Ravine is of horseshoe shape, and the opposite outer cliff is more than a thousand feet in height. The bottom slopes upward towards the backward crescent wall, the rim being quite level. The path, which is plainly marked by splashes of white paint, runs along the centre of the basin by the bed of a stream. In front is seen the grand front of the sheer precipice, lying some distance off and up under the summit of Mt. Washington. This symmetrical wall has been called the Mountain Coliseum. Its back wall, unless the season is very dry, glitters with innumerable streams of water called the Thousand Streams. From the base of this great wall a tortuous and difficult path continues up the mountain side to the summit, a distance of about a mile. The Snow Arch is one of the interesting features in early summer. [See Snow Arch.]

TUMBLE-DOWN DICK. A small mountain northeast

from Copple Crown Mountain [see Copple Crown], more easily ascended, and affording a similar view.

TWIN MOUNTAINS. Two prominent peaks of the Franconia range, 4920 feet high. A path has been constructed under the auspices of the Appalachian Mountain Club to the summit of the principal of these peaks, which runs from the Twin Mountain House. The view is one of the finest to be had in the mountains. The range rises from the southern bank of the Ammonoosuc River, and, running at right angles to the stream, stretches to the southward in the direction of the eastern bank of the Pemigewasset River. Its principal members are the North and South Twin and Mts. Guyot and Bond. Only about 200 feet depression separates the summits of the Twins.

TWIN MOUNTAIN HOUSE. A station of the Concord & Montreal and Maine Central railroads, 202 miles from Boston, surrounded by mountains. A view is obtained here of the great White Mountain Range. Close by are the Twin Mountains. [See Twin Mountains.] The mountain peaks visible are the Baby Twins, Mt. Hale, the North Twin hiding the summit of its southern brother, Mts. Garfield, Lafayette, Cleveland, and Agassiz.

U – V

UMBAGOG LAKE. See Rangeley Lake.
UNITOGA LAKE. See Newport, N. H.
UNITOGA SPRINGS. See Newport, N. H.
UPPER BARTLETT. See Crawford Notch.
VINEYARD HAVEN. See Martha's Vineyard.

W

WALKER'S FALLS. A beautiful fall of water in the Franconia Notch [see Franconia Notch], about two miles from its southerly extremity, and half a mile from the main road. It is reached by a path up the banks of a brook which crosses the road. The quantity of water is never very large, but it comes leaping down over a regular succession of stone steps, extending across the whole breadth of the bed of the stream, for a distance of about 30 feet. Half a mile farther on there is a larger and more picturesque fall, where the water descends at one leap a distance of about 60 feet. The sides of the brook are of precipitous rocks, somewhat resembling those of the Flume. [See The Flume.]

WARREN, N. H.   A mountainous town, a station of the Concord & Montreal Railroad, 146 miles from Boston, in which there are said to be more than 100 trout brooks.   The most picturesque of these is Hurricane Brook, which flows from Mt. Carr. In the vicinity is a deep gorge called Jobildunk Ravine, east of Moosilauke. [See Moosilauke.] In it are some beautiful cascades, which, owing to the difficulty of access, are seldom visited.

WARREN SUMMIT.   A station of the Concord & Montreal Railroad, 115 miles from Boston.   It is 1063 feet above the level of the sea, and the highest point upon the main road.   The view from here is remarkably comprehensive and beautiful.   Near the summit the train passes through a rock-cutting, three-quarters of a mile long, and, in places, 60 feet deep.   This work required the labor of 150 men for a year and a half, and cost over $150,000. As the descent toward the Connecticut Valley is commenced, the bold cliffs of Owl's Head are seen upon the right.

WASHINGTON LYING IN STATE.   An idea suggested by the appearance of the peaks on the east side of the Franconia Notch; Mt. Liberty serving as the face with its highest ridge as the nose.

WATERBURY, VT., reached by the Central Vermont Railroad, 202 miles from Boston and seven miles below Montpelier, is a popular resort for tourists, being in the immediate vicinity of Mt. Mansfield [see Mt. Mansfield], Camel's Hump, Bolton's Falls, and other places of interest.

WEIRS, THE.   A station on the Concord & Montreal Railroad, 109 miles from Boston.   It is situated on the shore of Lake Winnipesaukee [see Lake Winnipesaukee] with a fine, and comprehensive outlook across its waters upon the lofty mountains beyond. In colonial times the Indians had fish weirs here, in the shallows near the outlet of the lake, and caught shad.   It is from this fact that it derives its name.   Once a year the Winnipesaukee tribe encamped here, and passed weeks in feasting.   It was, until within a few years, little more than a boat landing, but its charming location and excellent railroad and steamboat facilities attracted the attention of various societies, and it has become a famous resort where different organizations hold their summer meetings and conventions. The most prominent of these are the New Hampshire Methodists and Unitarians, who have erected a small village in a grove a short distance from the landing, where they hold their camp-meetings, and the New Hampshire Veterans' Association, an organization which annually holds a large out-of-door gathering, and owns a number of buildings. There are, on each side of the railroad, capacious groves where summer meetings of various kinds are held, groups of cottages, and the office of "Calvert's Weirs Times," a paper published during the summer season by Matthew H. Calvert. In the view from the Weirs the Ossipee Mountains are seen to excellent advantage. On their left is to be seen the sharp peak of Chocorua. Red Hill is also visible, and stretching off towards the left are Paugus, Passaconaway, Tripyramid, Whiteface and Sandwich Dome. By going a little distance above the Weirs, a view of Mt. Lafayette can be had.

WELLS BEACH, reached by stages from Wells, Me., six miles, is a great rendezvous for sportsmen, its six miles of beach being frequented by snipe and curlew. A large trout stream crosses the beach, and in the woods partridges and woodcock are abundant. The views in the vicinity are particularly fine.

WHITEFIELD, N. H. A town on the Whitefield and Jefferson Branch of the Concord & Montreal Railroad, 200 miles from Boston. From Kimball and Highland Hills fine views are obtained of both the White and Franconia ranges. Mt. Garfield, formerly known as the Haystack, and several other high peaks.

WHITE ISLAND. See Isles of Shoals.

WHITE MOUNTAIN NOTCH. See Crawford Notch.

WHITE MOUNTAIN RANGE. While the whole mountain region of New Hampshire is about 40 miles square, the name "White Mountains" is popularly applied, for the sake of distinction, only to the great range which extends from the Crawford Notch [see Crawford Notch] northeasterly to Mt. Madison, a distance of 14 miles, with Mt. Washington for its culminating point. The mountains, beginning at the Notch, are in the following order: Webster, Jackson, Clinton, Pleasant, Franklin, Monroe, Washington, Clay, Jefferson, Adams, and Madison. The earliest printed account of the

White Mountains appears in John Joselyn's "New England Rarities Discovered," published in 1672, in which he says: "Four score miles (upon a direct line) to the Northwest of Scarborow a Ridge of Mountains run Northwest and Northeast an hundred leagues, known by the name of the White Mountains, upon which lieth Snow all the year, and is a landmark twenty miles off at Sea. It is rising ground from the Sea Shore to these Hills, and they are inaccessible but by Gullies which the dissolved Snow hath made; in these Gullies grow Saven Bushes, which being taken hold of are a good help to the climbing Discoverer. . . . . The country beyond these Hills Northward is daunting terrible, being full of rocky hills, as thick as Molehills in a Meadow, and cloathed with infinite thick woods." The Indian name of the great range was Waumbek Methna (mountains with snowy foreheads).

WILDCAT MOUNTAIN. See Jackson.

WILLEY HOUSE. See Crawford Notch.

WILLOUGHBY LAKE, situated in the town of Westmore, Vt., is reached by stage, eight miles from Barton's Landing, a station on the Passumpsic Division of the Boston & Maine Railroad, nine miles from Newport. It is of a crescent shape, seven miles long, and from a half to two miles wide, and lies between two mountain peaks, with nearly perpendicular sides, whose bases apparently meet far below its surface. The depth of the lake is a matter of conjecture, a sounding-line of 700 feet having failed to touch bottom. The mountains rise so abruptly from the shores that there is room for little more than a carriage way around its margin. The mountain on the east side is Annanance or Pisgah, and is 2638 feet high. That on the west side is Mt. Hor, and is 1500 feet high. The summit of Mt. Annanance is reached by an easy bridle path, two miles long, and from it is a fine view including Lake Memphremagog and Owl's Head on the north, Lake Champlain on the west, the entire range of the Green Mountains, and every prominent peak of the White Mountain group. Other places of interest are the Devil's Den, a dismal hole on the lake shore, at the base of a perpendicular precipice 600 feet high, and the Flower Garden, a spot on its summit.

WING ROAD, N. H. A station of the White Mountain Branch of the Concord & Montreal Railroad, 194 miles from Boston, where the White Mountain trains leave the main line and continue up the Ammonoosuc Valley. The station is within the limits of the town of Bethlehem, and from it a fine view of Mt. Lafayette and Twin Mountains is had.

WINNEWETAH CASCADE. See Jackson.

WOLFEBORO, a village named in honor of General Wolfe, the hero of Quebec, is beautifully situated at the head of a bay, and is one of the most attractive points on Lake Winnipesaukee. The views in the vicinity include lake and mountain scenery in infinite variety; and the visitor may indulge to his heart's content in the delightful recreation of mountain climbing without going far from his hotel. Among the numerous excursions which can be taken from this place, one of the most interesting is to Wentworth House, on the borders of Smith's Pond, the summer residence of the colonial governor Wentworth of Portsmouth. Boating forms a favorite amusement with summer visitors at Wolfboro, the calm, placid waters of the lake affording excellent opportunities for this diversion. A branch of the Northern Division of the Boston & Maine Railroad, about 13 miles long, connects with the main road at Wolfboro Junction for North Conway, the time between the two principal points being but one hour.

Y

YORK BEACH, near the quiet little village of York, N. H., nine miles northeast of Portsmouth, is a popular summer resort, reached by steamer or stage. The beach slopes gently to the water from the eminences behind, and affords excellent bathing. There is also good fishing in the vicinity. Cape Neddick runs out into the sea at the north end of the beach, and a short distance inland is Mt. Agamenticus, from the summit of which are to be had fine views of the White Mountains, the ocean, and the harbors of Portsmouth and Portland. Bald Head Cliff is a remarkable promontory five miles north of York Beach, of peculiar formation and affording fine views. Beyond, stretching away to Wells Beach [see Wells Beach], is the long Ogunquit Beach.

YOUNG MAN OF THE MOUNTAIN. See Crawford Notch.

All of the . . .

Principal New England

. . . Summer Resorts

ARE DIRECTLY REACHED BY THE

Boston & Maine Railroad,

AND ITS CONNECTING LINES.

THE WHITE MOUNTAINS,
GREEN MOUNTAINS,
ADIRONDACK MOUNTAINS,
AND MOUNT KINEO.

The North Shore, Isles of Shoals, York Beach, Kennebunkport, Old Orchard Beach, Bar Harbor, and St. Andrews, N. B.

Lakes Winnipiseogee, Champlain, Sunapee, Megantic, Moosehead, and Rangeley.

Summer Publications.

Complete list of books descriptive of all summer resorts, also summer excursion book, giving list of hotels and boarding houses, excursion rates, and much other valuable information, will be mailed free upon request. Address General Passenger Department, Boston & Maine Railroad, Boston.

D. J. FLANDERS,
G. P. & T. A.

THE . . .

Maine Central Railroad

IS THE

· · · CONNECTING LINK

FOR ALL PLACES NAMED IN THIS PUBLICATION,
EITHER DIRECT OR BY ITS RAIL LINES,
OR IN CONNECTION WITH ITS
STEAMER LINE THE

Portland, Mt. Desert, and Machias Steamboat Co.

Express trains daily to all parts of Maine. Pullman Sleeping
and Parlor Cars, extra charge, — regular coaches, no extra
charge; all vestibuled, lighted by gas, running over a
perfect roadbed. Its Schenectady and Rhode Island locomo-
tives the heaviest taking water from track tanks.

We publish Maps and Guides giving full information
about all the large and small resorts, hotels, boarding and
farm houses.

Send for our list of publications.

F. E. BOOTHBY,
Gen'l Pass. Agent.

PAYSON TUCKER,
V. P. and Gen'l Manager.

Kingdon's Dictionary

OF THE

White Mountains

AND OTHER

New England Summer Resorts.

A concise, convenient, and comprehensive book of Reference for Tourist and Traveller.

Full and accurate information regarding places and objects of interest in New England and how to reach them.

Single copies mailed to any address on receipt of twenty-five cents. ._ .

ADDRESS **S. S. KINGDON,**
BOSTON, MASS.

www.ingramcontent.com/pod-product-compliance
Lightning Source LLC
Chambersburg PA
CBHW020335090426
42735CB00009B/1546